Creative

Think Differently

Inspirations

Dreams

Great

Make A Difference

Msafiri V.G Lwihula

I0052497

Think

Differently
Make A Difference

"Change The Game"

TABLE OF CONTENTS

FOREWORD

This book is about creating awareness and inspiring to the Tanzanian and African youth generally to be creative, innovative and inventive.

I will take a deep insight on different angles from scientific, social, governmental, cultural and eating traditions that helps someone out there to boost his brain innovative ability .I will also hint on safer ways for inventors to follow and protect their work.

Why creation, innovation, invention.?

The Simple answer is the today's world is a digital world and not analogue anymore. Where does all this come from?

The computers, internet, mobile phones, advanced science in health and military, social networks, satellites and exploration of earth and outer space, mining and all other type of scientific activities and gadget you see on earth today. All of these did not come just like magic. Not at all, some human being out there did their best to make it all happen; on their own effort but undoubtedly with support from government and communities.

This book will help you as a child, parent, innovator, entrepreneur, teacher and a government leader to see how we can put our efforts and mind together to make a difference in creating a new Tanzania in a new Africa

ABOUT THE AUTHOR

I was born on January 26, 1986 at St. Francis Hospital Ifakara (Kilombero-Morogoro), My Mother was a nurse at St. Francis Hospital and My father was a lecturer at Muhimbili University of Health and Applied Science,(Both deceased 1996,2003-Rest In Peace).

I went to primary school at Mapinduzi Primary school in Morogoro-Ifakara till standard five when I shifted to Arusha, Where I joined Meru Primary School Completing standard seven there.

After that I joined Forest Hill Secondary school (Morogoro-Town) for some time and later transferred to Panda Hill Secondary School Where I Completed my O-level education (2004). I later joined Shaban Robert Secondary school (Dar es Salaam) for High school and later transferred to Msangani High School (Kibaha-Pwani) where I Completed My A-level Education (2007)

In 2008 I joined Institute of Finance Management (IFM) for Higher education where I studied Bachelor of Science in Information Technology and graduated in the year 2011. Currently I am working as IT specialist at Fabec Investment Limited (Dar es Salaam), I am also a writer, Innovator, Idealist, and entrepreneur.

Msafiri V.G Lwihula

ACKNOWLEDGEMENT

I would like to express my gratitude to the many people who saw me through this book; to all those who provided support, talked things over, read, wrote, offered comments, allowed me to quote their remarks ,share their stories and assisted in the editing, proofreading and design.

I would like to thank the Almighty God, for life, love and blessings that enabling me to publish this book. Above all I want to thank my Father and mother (Late Prof George K. Lwihula, ELisaria Mrema),My Guardian Mr Adolph K.R Banyenza, my sisters Flora Mrema and Neema Alinda George and the rest of my family, who supported and encouraged me through the whole process of making this book possible.

I would like to thank Mrs. Sekita Nsekela, Mr. Salim R Sultan,Mr. Frank Munisi and Mr.Orton Kiishweko for helping me in the process of selection and editing. Thanks to Mr. Laurent Makole my publisher who encouraged me from the beginning to the end.

Thanks to Mr.Christopher Kamugisha,Mr.Lumuli Mwakoba, and Mr Haji Hamza(GRM) for book cover, drawings and images - without you this book would never have had good cover and images, without forgetting Mrs Doreen Anthony Sinare (HEAD OF COSOTA) and Engineer Joel Makyao for their continuous support and encouragements.

Last and not least: I beg forgiveness of all those who have been with me over the course of the years and whose names I have failed to mention.

CONTACTS

Msafiri Vicent George Lwihula
P.O.Box 866
Dar es Salaam
Email: msafirilwihula@gmail.com
vincirra@gmail.com

"Future depends on what we do now"

Think

Differently
Make A Difference

"Change The Game"

CHAPTER ONE

"ALL YOUTHS ARE POTENTIAL INNOVATORS"

"ALL YOUTHS ARE POTENTIAL INNOVATORS"

INNOVATION

Is the process of translating an idea into a good invention or service that creates value or for which customers will pay and is when an inventor becomes entrepreneur. Or

Innovation is the process of making changes to something established by introducing something new that adds value to customers (O'Sullivan), 2008.

To be called an innovation, an idea must be replicable at an economical cost and must satisfy a specific need. Innovation involves deliberate application of information, imagination and initiative in deriving greater or different values from resources, and includes all processes by which new ideas are generated and converted into useful products. In business, innovation often results when ideas are applied by the company in order to further satisfy the needs and expectations of the customers. In a social context, innovation helps create new methods for alliance creation, joint venturing, flexible work hours, and creation of buyers' purchasing power. Innovations are divided into two broad categories:

(A) Evolutionary innovations (continuous or dynamic evolutionary innovation) that are brought about by many incremental (continuously) advances in technology or process.

(B) Revolutionary innovations (also called discontinuous) which are often disruptive and new.

Innovation is synonymous (associated) with risk-taking and organizations that create revolutionary products or technologies take on the greatest risk because they create new markets.

Imitators take less risk because they will start with an innovator's product and take a more effective approach. Examples are IBM with its PC against Apple Computer, Compaq with its cheaper PC's against IBM, and Dell with its still-cheaper clones against Compaq.

In our country imitators example can be of Money transfer systems, Whereby MPESA system started earlier faced a lot of challenges while other systems which followed later operated in a little bit lower risk compared to MPESA because at least they had a clue of the business environment they are getting into.

When we talk about innovation there are other two concepts associated with it. We should discuss them before going further so as to understand the whole concept better. These concepts include CREATIVITY, CREATION and INVENTION.

CREATION, CREATIVITY

There are many ways of describing creativity here are some of them,

-Creation is a phenomenon whereby something new and valuable is created (such as an idea, a joke an artistic or literally work, a painting, a musical composition, a solution an invention etc)-(Wikipedia)

The ideas and concepts so conceived can then manifest themselves in any number of ways but most often into something we can see, smell, hear, touch or taste. The range of scholarly interest in creativity includes a multitude of definitions and approaches involving several disciplines; psychology, cognitive science, education, philosophy ,science, technology, theology, sociology, linguistics, business, studies, songwriting and economics; taking in the relationship between creativity to general intelligence, mental and neurological processes associated with creativity, the relationships between personality type and creative ability and between creativity and mental health, the potential for fostering creativity through education and training, especially as augmented by technology and the application of creative resources to improve the effectiveness of learning and teaching processes.

Example: Computer we are using in today's world they are product of creativity, all of them were in someone's mind but after a while it became possible to turn the idea/vision into a product.

SIMPLE ART OF CREATION

By nature human beings are problem solver, in our daily lives from sunrise till sunset our minds are busy solving life issues, if not about food, it could be environment, health, education, social problems or political matters.

The ability of human being to solve simple/big problems that his faces around his environment is creativity, before even getting to the big creations and technology; let's use normal and common life examples,

A MATCH AND MATCH BOX

Our history tells us that our ancestors where facing problem of cold weather, hostile animals and also shortage of food. That is when people thought of something like fire, through the period of time fire sources have been developed but the simple one and which we are using daily is a match. This helps us in starting fire at our households for cooking, burning bushes and all other things related to that.

Match box

Small Kerosene lamp

Humans saw the need of light at night especially for night vision, security and decoration they develop kerosene lamps.

These are simple creations with great use in our daily lives especially in our African countries where majority of people still have no access to electricity. Most of Africans have used these kinds of lamps to read at night at our schools and households, so it's among the inventions which have major effects in our lives.

KNIFE

A knife is an instrument for cutting, it can be used for cutting food, also can be used as protective gear, for hunting and agricultural tool.

Still focusing on our daily lives simple creations that are affect our lives, let's have a look at the art of cooking.

Food

COOKING ART

Cooking is an art and creativity. The variety of delicious foods we eat in our daily lives from breakfast, lunch to dinner they are all been developed by people. People from different areas have developed different recipes due to the availability of certain types of products in their areas. Important to note that human creativity was used here.

African foods like UGALI, WALI, MAKANDE, NDIZI, and VIAZI among others. At first people didn't know if such varieties were really good food but with time, things changed and they started using them. We are learning that our ancestors used to eat fruits and roots of some trees, but with time people tried a lot and came up with different tastes. Due to scarcity of some food varieties/resource; if there are different varieties of food recipes it helps in minimizing the usage of one food source/material which might later diminish.

Creativity also can be seen on this level when you look at the way one material can produce so many things. Example rice, it can be cooked as rice, can be used for porridge, also pilau, birian; same applies to Maize and wheat flour.

So the creativity concept is a wide one which touches different angles of life. The simplification of how things are done, for example people were struggling to carry water from one place to the other and over time, they came up with buckets and the likes to solve the problem.

I am emphasizing the use of innovation and creativity in our daily lives to solve our daily problems and that's where we see how important this concept is. We will respect the creations of others and we will be able to prosper as a result. If you're living in a place/village/ home where you're facing some types of challenges, it's your duty to think about something to do in order to solve that.

If there is a problem of water in the village, people should think of the ways to simply get water for themselves even if government didn't do anything about it, be it taping from the river somewhere or dig a well.

There is a story I saw on television (ITV) sometime back of one person in Iringa region who created a local water system for his village; simply by the use of bamboo trees as water pipes. He was able to tap water from a source and distributed to himself and others in the village. These are kind of people we need in our society. Solution minded people.

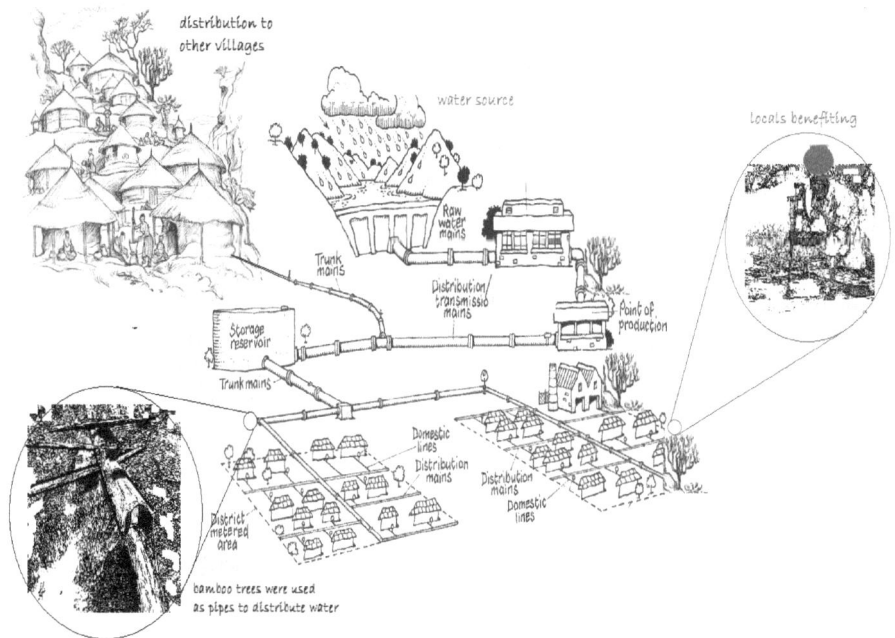

distribution to other villages
water source
locals benefitting
Raw water mains
Trunk mains
Distribution/ transmission mains
Point of production
Storage reservoir
Trunk mains
Domestic lines
Distribution mains
Distribution mains
Domestic lines
District metered area
bamboo trees were used as pipes to distribute water

To stimulate creativity, one must develop the childlike inclination for play and the childlike desire for recognition.-Albert Einstein

Prompted by initiatives like these. I am emphasizing people to face their challenges around areas and act upon them; be it power, school, health services and all other things citizens must act first. This way will promote innovation/creativity and development as a whole. I believe when something new and good is discovered even God up in heaven smiles because we have used our mind he gave to us good enough.

Let's look at another simple African life experience on our streets. There are times during rain seasons where there are a lot of rains in a week or in months but as soon as rain ends, immediately thereafter a problem of water arises. For years now, we don't use our resources well, in our household we don't even try to find out how to tap and reserve rain water for use in dry season for various purposes. I don't believe that we are not capable of coming up with simple solutions but what I know and believe is that we don't sit down and think about situations and find answers. Most of us we are comfortable with our problems. For a human being this is a bad habit. We need to think outside the box, taking advantage of situations and derive benefit from them.

This problem grows to government level. For years we have been depending on one way of generating electricity in our country which is hydroelectric (Water). Soon after rain stops, the country starts facing power outages and black out; while we have number of other sources from Wind, gas, thermal and even nuclear resources. These can produce enough energy for the whole country and for sell outside, but why are we failing? There are number of reasons. Major reasons is, we Tanzanians we don't have the tradition of looking at things from different angles. We don't use our brains good enough to think about solutions even before the problem occurs. The habit should be injected into our families' level (households) and schools to young blood. If we teach them that they need to be proactive on situations and not to be complacent, these problems will be very much minimized in the future or non-existent. They need to be problem solvers and not blaming, they should always think and predict some things and find out solutions before they happen.

A CASE STUDY OF TANZANIAN CAPITAL CITY DAR ES SALAAM.

Most of Dar es Salaam streets are facing floods problems during rain seasons at places like Sinza, Kinondoni, Jangwani, Msasani, Mwenge. It's been years and this problem is still there; but what I see for us as citizen and government in general has failed to do something about it. I can easily conclude that we are not innovative at all because if we were this problem could not have been there anymore.

The problem of drainage systems starts from our streets. Poor land planning is in our government system, management of wastes are on our streets and municipal hands but we act like we don't see. We all blame when things go wrong during rain seasons. There is a huge part that needs to be played by citizens on these streets with government support. If we all act reasonably enough, there would be no floods on our streets. This is an alert to us all, that we need to use our mind and creativity to solve problems surrounding us. And not wait for someone out there to come and solve for us. People must know that in order for us to grow, we need to be responsible in problems solving in our respective areas. People should share and utilize their vision, unique thoughts/ideas to help the community; in return the community will accord them respect and lifetime remembrance plus embracement.

Image shows Dar es Salaam floods which happens every year.

INVENTION

It's when new scientific or technical idea, and the means of its embodiment or accomplishment (Simply it is a introduction of something very new to the world). To be patentable, an invention must be novel, have utility, and be non-obvious.

To be called an invention, an idea only needs to be proven as workable. But to be called an innovation, it must also be replicable at an economical cost, and must satisfy a specific need. That's why only a few inventions lead to innovations because not all of them are economically feasible.

What is the difference between innovation and research and development (R&D)?

According to Baum, (2004), CEO of Dial, R&D means improvements to existing products, quality assurance, line extensions. And Innovation simply is something truly different in the market that makes your customers lives better.

ORGANIZATIONAL INNOVATION

Prather (2010), innovation is a social process requiring an effective team to bring a good idea to fruition in the market place.

Prather's (2010), experience leading the DuPont Center for creativity and Innovation led him to develop the innovative Competence model consisting of the three arenas of education, application and leadership. Developing an internal competence for innovation requires a systemic approach in all three arenas.

DuPont Initially concentrated on the education and application arenas and naively assumed that the leaders would know how to lead it. 'We know now that total

Leadership commitment from the top is the single most important factor in a company's level of innovation competence and its innovation success' (Prather

2010).

It is people, not organizations, who come up with the innovative ideas. This report is a review of the Literature on organizational and management innovation. The discussions begin with a systems model of the innovation value chain, which is followed by discussion of internal factors such as management Innovation, leadership, organizational culture, organizational design and human resource management.

The discussion then moves to innovation involving interaction between the organization and the external environment. External interactions include partnerships between R&D organizations and businesses.

Interactions between consumer/customer and researcher, open innovation strategies and the recent international development of an Indigenous entrepreneurship research paradigm. The paper concludes with a model for implementing an innovation framework within a public-good research organization.

INNOVATION VALUE CHAIN

Hansen & Birkinshaw (2007) maintain that there is no universal solution for improving innovation in organisations. They believe that management needs to take an end-to-end view of their innovation efforts, pinpoint their particular weaknesses, and tailor innovation best practices as appropriate to address the deficiencies.

The innovation value chain offers a framework that breaks innovation down into three phases (idea generation, conversion, and diffusion) and six critical activities (internal, cross-unit, and external sourcing, idea Selection and development; and spread of the idea) performed across those phases.

Hamel & Getz (2004) believe that companies need to institutionalize innovation as a core value. They also believe that ideas can come from anywhere within the company, not just R&D personnel. Consequently it is necessary to 'free your innovators'. Companies, such as Gore, that have totally embraced an innovation policy have moved into very diverse areas, as well as being one of America's most highly rated employers

(Hamel & Getz 2004).

	IDEA GENERATION			CONVERSION		DIFFUSION
	IN-HOUSE	CROSS-POLLINATION	EXTERNAL	SELECTION	DEVELOPMENT	SPREAD
	Creation within a unit	Collaboration across units	Collaboration with parties outside the firm	Screening and initial funding	Movement from idea to first result	Dissemination across the organization
KEY QUESTIONS	Do people in our unit create good ideas on their own?	Do we create good ideas by working across the company?	Do we source enough good ideas from outside the firm?	Are we good at screening and funding new ideas?	Are we good at turning ideas into viable products, businesses, and best practices?	Are we good at diffusing developed ideas across the company?
KEY PERFORMANCE INDICATORS	Number of high-quality ideas generated within a unit.	Number of high-quality ideas generated across units.	Number of high-quality ideas generated from outside the firm.	Percentage of all ideas generated that end up being selected and funded.	Percentage of funded ideas that lead to revenues; number of months to first sale.	Percentage of penetration in desired markets, channels, customer groups; number of months to full diffusion.

(A)INNOVATIVE CASE STUDY (WORLD WIDE)

(a)Ninti One Limited

CRC-REP Working Paper CW001

Organizational Innovation: A review of the literature Ninti One Limited

Another American company, Dial, has embarked on an innovation campaign that extends from scientist to secretaries. Innovation is now a core value and the company encourages and rewards innovative employees. Firstly, they changed the name of the R&D Center to the Center for Innovation and created a Vice President of Innovation who reports directly to the CEO. To encourage internal innovation, the company holds an annual innovation day where all employees can learn about creativity and generate

Ideas. Employees are given awards for innovation and scientists are required to develop two new patentable ideas per year. Dial also solicits outside ideas through a competition that is published through inventors' organizations and publications (Baum 2004).

However, developing an innovative culture is much more than offering a few rewards for innovative behavior. For example, Thompson & Heron's (2006) research looked at links between three dimensions of the employment relationship – the psychological contract, affective commitment and knowledge-sharing behaviors – and their consequences for innovative performance. The results suggest that organizations

that invest heavily in socialization of employees, and enact policies and practices to forge strong personal identification with the organization and its values and purpose may be better placed to appropriate value from knowledge worker behavior.

Newman (2009) developed the acronym CREATIVE to describe the elements needed to develop an innovative R&D culture. CREATIVE stands for Customer-focused, Risk-tolerant, Entrepreneurial, and Alignment with strategy, Technology and scientific excellence, Innovative, Virtual organization (Collaboration) and Execution.

Establishing innovative culture is a challenging, long-term task and is not achieved through a single seminar, tool or program. There are no simple techniques that can create this culture and make changes in intrinsic values.

The culture of an organization is also shaped by its design. For example, bureaucracies and hierarchies often work to suppress creativity and innovation and therefore may not be conducive to fostering an innovative culture within the organization.

ORGANIZATIONAL DESIGN

Organizational structures that may have worked well in the past are not suitable when an organization's goal is to be innovative and creative. Bureaucracy generally sets an upper limit to what employees are allowed to do in their work, which inhibits creativity and innovation. Organizations that have achieved innovation have changed their organizational design depending on the nature of the industry and their own goals.

Garner (2008) believes that it is necessary to return power to the researchers by reorganizing R&D into small, highly focused groups headed by people who are leaders in their scientific fields and can guide and inspire their teams to achieve greatness. For example, at GSK the organizational pyramid was broken into a constellation of highly focused centre of excellence designed to improve transparency, increase the Speed of decision making, and restore freedom of action to the scientists actually conducting the research.

American company Gore has virtually no hierarchy and associates (employees) can spend up to 10% of their time dreaming up new applications for the company's unique products. When an idea emerges the Innovator has to recruit colleagues to support its development. Gore's democratic innovation policy has seen the company move into very diverse areas, as well as being one of America's most highly rated.

(b)Wal-Mart Company-(from www.moneynewsnow.com)

Wal-Mart Company is one the largest store in USA and in the world, having about 8500 stores in 15 countries with 15 different names, they are also considered as biggest retailers in the world, and largest grocery in the United States of America. It is incredible to think of exactly how big this company is, but they couldn't have gotten this big without the proper creative/innovative marketing strategies, so let's see how they thought differently with their tactics and see if we can learn a thing;

(i)Low prices

The first thing anyone notices when going into a Wal-Mart is incredibly low prices they offer their products, for this itself helps with their marketing strategy because everyone is always looking for the lowest price and it can usually be found at Wal-Mart. Once someone sees an advertisement about low prices and actually visits a store or their websites; they will become a loyal customer since they will be able to find the best deals all year long.

(ii) Easy accessibility

One of the main things about Wal-Mart is that they are extremely accessible, they have thousands of stores all around and if you don't live near one, you could always use their internet store all around and buy from there, being this accessible makes marketing to people much easier since the customers doesn't need to work hard to do business with them. Customers are also able to contact them 24 hours a day, which helps with the customer satisfaction aspect of the whole business.

(iii)Appeal to as many people as possible

Wal-Mart offers products of all types, from beauty products, home cleaning products, toys, electronics, to even groceries, anyone could find something they have a use for there, which gives them a much broader appeal to all personality types, although most of us won't be providing as many product types as Wal-Mart, it is good to keep in mind that by expanding your business and putting new products or services for sale you can increase your overall customer base.

(iv) Social media campaigns

All large companies including Wal-Mart, have focused a lot of resources on social media campaigns, this is because it is one of the best types of marketing due to its consistent results. There are many different social media strategies that are out there, but the two that have proven to be the best would be blogging and video sharing and then linking those content posts with your social networking profiles.

Video sharing sites such as YouTube, gets millions of hits every day, so why not try to direct some of those people to your company? If you make a relevant and quality video that will appeal to your target audience and at the same time promote your products or services you could see your revenues spike due to increased of sales.

(c)Amazon.com (from marketingplan.net)

Jeff bezos, Amazon.com founder and CEO, dreamed about books, in 1994, he created Amazon.com inc, which he labeled as "Earth's Biggest Bookstore." The ecommerce company went online in 1995 and soon expanded into other media, including DVDs, VHS, and CDs, MP3s and eventually a wide range of other products, including toys, electronics, furniture and apparel. As such, the taline soon changed to "Earth's Largest Selections." But books were only beginning of Bezo's up and coming enterprise.

Amazon.com went public in 1997. In the first shareholder letter, Bezos penned the fundamental foundation for Amazon.coms success."Start with customers, and work backwards, listen to customers, but don't just listen to customers- also invent on their behalf (think different).. Obsess over customers." This policy was backed by a starting business philosophy-Bezos planned on operating at a loss for 4-5 years; it was not until 2001, that Amazon.com claimed over 1. 4 million customers after only two years of being online.

Now, 45 million satisfied customers shop at Amazzon.com for everything from books (Most popular) to fashion apparel to fine jewelry to Christmas toys. It has one of the most recognized brand names in the world and garners an estimated 50% of its sales from overseas consumers. Surviving the dot.com bust of the late 1990s and early 2000s. Amazon weathered the e-storms and now thrives in the retail and centered on customer fulfillment. Amazon.com proceeds into the next decade with a profit firmly in one hand and the capacity to blow it out of the water in the other hand.

AMAZON.COM BUSINESS PHILOSOPHY.

Despite its massive growth, Amazon.com remains unremittingly focused on the customer, out of 452 company goals in 2009, 360 directly affected customer's experience.

Amazon.com's self-proclaimed mission stament is "We seek to earth's most customer-centric company for the primary customer sets: consumer customers, seller customers and developer customers."

In a special for the Miami Herald, Journalist Jack hardly declares." Customer obsession; Innovation; bias for action; ownership; high hiring bar and frugality. These six core values focus Amazon.com operational strategies." It is committed to long-term growth based on consumer satisfaction.

Amazon.com bases its marketing stratagem on six pillars.

(i) It freely proffers products and services

(ii) It uses a customer-friendly interface

(iii) It scales easily from small to large

(iv) It exploits its affiliates products and resources

(v) It uses existing communication systems

(vi) It utilizes universal behaviors and mentalities.

Much of its marketing is subliminal or indirect-it does not run $1million dollar ads during super bowls nor post flyers in mall marketplaces. Amazon relies on wily online ploys, strong partner relations and a constant declaration of quality to market itself to the masses.

(d)Virgin Atlantic airlines-(bussinesscasestudies.co.uk)

Anybody who has bought a car will tell you that there are tangible points of difference which affect their decision to make purchases. For example, there are many different models of car which have various characteristics. They may have manual or automatic transmissions, fuel injection systems, be small and versatility designed for family or for couple, have four-wheel drive and also have vast.

In airline industry Virgin Atlantic has set the standard for excellent customer care. It has introduced a number of unique and innovative products to the business, which have become not only the hallmark of virgin's services values, but also of the airline industry as a whole, for example, virgin Atlantic was the first airline to offer just two flight classes, economy (a choices of premium).

Virgin Atlantic has successfully used branding to develop perceptions of a corporate personality, offering a unique combination of services elements and attributes, which serve to set themselves apart from the competition.

In other words they have developed a "service brand" symbolizing both individualistic attributes and quality of service. This means that even customers with limited budget can enjoy the flight.

(e)Invention of bulb-(Wikipedia)

Electric light

Thomas Edison is the person who is credited as a bulb inventor, but in the real sense Edison did not invent the first electric bulb, but instead invented the first commercial practical incandescent light, he thought differently and be able to turn the bulb invention to business (innovation).(1879)

Many earlier inventors had previously work on this including Alessandro Volta (1800),Humphrey Davy, James bowman Lindsay ,Moses G. farmer, William E. Sawyer, Joseph Swan and enrich Gobel. Some of these early bulbs had such flaws as an extremely short life, high expense to produce, and high electric current drawn, making them difficult to apply on a large scale commercially.

After many experiments, first with carbon filaments and then with platinum and other metals, in the end Edison returned to a carbon filament, the first successfully test was on October 22,1879, it lasted 12.5 hours.

Edison continued to improve this design and by November 4, 1879, filed for US patent 223,898(granted on January 27,1880) for electric lamp using "a carbon filament or strip coiled and connected to platina contact wires". Look at how useful this technology has turned out to be.

(B) Innovative Case Study (Domestic Examples)

(a)FAST JET AIRLINE COMPANY

Fast jet is an airline company which operates in different parts of the world including Tanzania. When they came into our market due to market competition from Precision Air and other airlines, they had to come up with a new business strategy of cheap flying cost by considering only the price of a seat per passenger; for the case of luggage and other services it is for the passenger to take them or not.

By this strategy we have seen them getting more customers. Now for example price of flying to Mwanza from Dar es Salaam is cheaper than going by bus. That way, they are attracting more customers and winning the market not only in Tanzania but they are among the fast growing airlines in East and Central Africa. Fast jet thought differently and they made a difference in the airline market in Tanzania.

(b) NMB (NATIONAL MICROFINANCE BANK OF TANZANIA)

One of the Tanzanian bank company that I salute for their capability of thinking differently (outside the box) and which has made a difference is National microfinance bank of Tanzania. For the past years, due to large number of customers it has, because they serve for most governmental employees, it's easy and cheap to open up an account; they found out that they had a lot of challenges to manage the customers especially ques in banks. The bank and customers suffered a lot from this, but thank to the development of mobile transfer service, due to the development of Telecommunication/ICT industry in Tanzania especially in the area of Money transfer services and mobile banking.

NMB embraces the technology very quickly and embedded it to their banking system so as to assist banking through mobile phones. By the use of mobile money transfer services like MPESA,TIGOPESA,AIRTEL MONEY,SELCOM,MAXMALIPO now customers are able to deposit and withdrawal money even without going to the bank. NMB did very good by embracing technology changes and also the Bank itself by promoting innovation in the organization acting quickly in order to solve problems and add value to the services they are providing. And this is thinking and acting completely differently.

(c)TIGO-MOBILE COMPANY TANZANIA

TIGO MOBILE COMPANY TANZANIA is among the successful telecommunication Company in the country, but market penetration in the country like this is not as easy as it seems. There is a lot of competitions from other telecom companies like Vodacom, Airtel, Zantel. So for the company to find its niche, it must come out with a great plan, company need to think differently.

Tigo specifically decided to focus on common Tanzanian citizens with low income, so their services were a little bit cheaper, and it was used by mostly ordinary/average Tanzanians mostly students.

It is perceived by many as the cheapest telecom company, which at the end of the day trapped so many Tanzanians. And look at how they are performing?. They are one of the leading telecom company in the Country. This is organization innovation.

(d)SMILE TELECOM COMPANY

This is the telecom company which came into the telecommunication market where there are number of other competitive companies around, so they studied the market and found out there is a lack of good/speed/reliable internet.

They decided to be very specific, focusing on internet service only, but they came differently. First of all they came with new internet/data technology of 4G LTE (i.e. FOURTH GENERATION LONG TERM EVOLUTION TECHNOLOGY) which is very high speed and reliable.

Now the outcome is awesome. They are providing the best internet in the country especially in areas like Dar es salaam and Arusha, where they started and are doing very well. Those are ways of thinking differently and strategies to penetrate the market.

(e)HTT-HELIOS TOWER TANZANIA

Mobile companies in the country wants to reduce the number of the activities that they control directly, they prefer outsourcing instead.

By doing this company reduces hassles and risk of dealing with many issues. Helios Company discovered that weakness; they turned the difficulties that are facing mobile companies into an opportunity.

Thus they decided to take the risk of handling all telecom towers in the country; they build towers from the ground to the end and also do the maintenance. The telecom companies only pay for the services.

(f) EMPLOYMENT AGENCY-(ERO LINK)

There has been a problem of finding workers and employees of different kinds, especially when they are needed immediately, be it professionals or casual laborers. It is always problems, challenges and difficulties which motivates clever people to think outside the box and come up with solutions by turning the difficulties into opportunities. Agency like ERO LINK Tanzania does a number of interviews to many job seekers; and holds their information in the database.

Hence whenever they are needed by a certain company, they communicate with them and link them with employers. In doing so agency got commission and the business continuous. This book is focuses on opening eyes to many of African youth to see the opportunities around us, wherever we are. If we try to think differently we will always find a way to survive and do something remarkable in our lives, society or even at workplace.

(g) AZAM/BAKHRESA COMPANY LIMITED

Azam Company limited is one of the biggest Tanzanian/East and Central African Company which deals with food products particularly for an ordinary citizen in the region. They thought very differently when they came up with their juice product.

They found out that in Tanzania there are a lot of seasonal fruits, such as mangoes, pineapples and oranges. During the season they are plenty and a price in the market gets low. Most of people just use them for normal and minimal use. We don't utilize them as they are supposed to.

So they decided to use these easy and not fully utilized resources to make different juices in a very high quality in the sense of hygiene and packaging which is most vital part of the process. And they put their products in the market in different varieties. Now they are doing big business.

Also on the side of coconut products like coconut milk, they thought differently on the matter, and came up with the best creative idea of preparing coconut milk and packaging it, for a simple use. They found out that the normal process of preparation is too long especially for the busy urban areas, people.

By doing this saves them a lot of time in food preparation and now the product is doing very well.

(h) NYAMA CHOMA FESTIVAL/INSTAGRAM PARTY

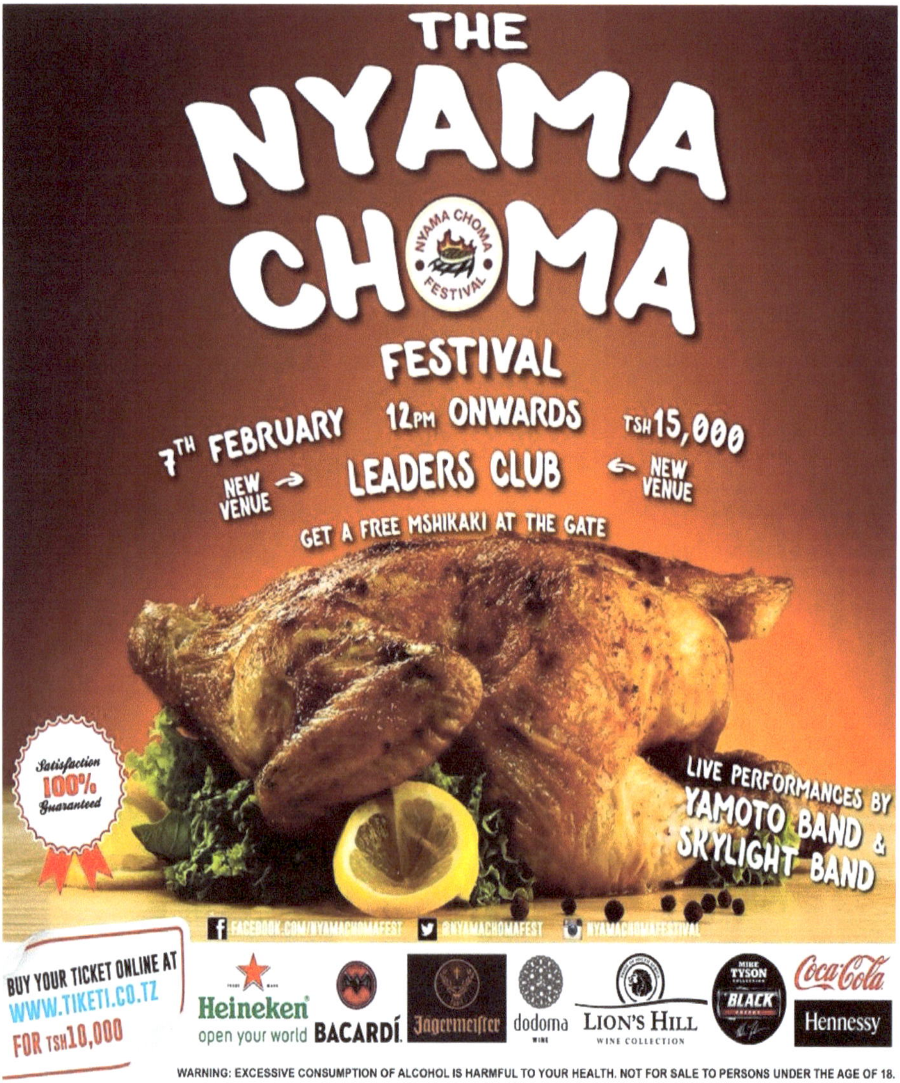

Miss Carol Ndosi thought differently/way beyond outside the box and she made a difference in her life and community in general by coming up with the idea of Nyama Festival in June 2011 which is a brand fully patented in Tanzania, Congo, and Uganda and soon will be in Rwanda.

An event which is occurring four times in a year, whereby there is a showcase of Nyama Choma/BBQ skills from different bar and people, people come and buy/eat and have fun, also there is entertainment(music and games).

Now it's becoming a tradition in different cities of Tanzania like Arusha, Mwanza, Mbeya, Iringa to attend her event and she is making some good money and helping others too.

Instagram is a social network site which is owned by Mark Zuckerberg the owner of facebook. Currently it is one of the busiest social networking site consisting many youths. Mr Freddy, Gillsant, Seth and others thought differently by using this opportunity of many youths to be in this social network, to start an Instagram party event where youth are getting together to have fun/relax and exchange ideas. This is what I call entrepreneur skills creativity, they turned a social networking into a business opportunity.

(j) BONGO MOVIES/BONGO FLAVA

There is no doubt that bongo movies/bongo flava has made a huge impact in helping youth in our community to get jobs. Most of Tanzanian youths are self employing and are also being employed by others in this industry.

I don't want to name many of them who played a great role and thought differently especially for taking it to the next level, but we have seen the huge effort of Late Steven Kanumba/Prof Jay/(Joseph Haule),Hon.Joseph Mbilinyi/Lady Jay Dee. Government is getting tax and artist and other stake holders are benefiting.

The industry is still small but it's growing very fast. Other major positive impact is the advertisement of our culture and growth of Swahili language. My suggestion on movies should be more original expression of real culture with a lot of creativity and diversity, for example we have stories of many heroes/leaders/superstars that are alive and others have passed away and we don't know much about their histories.

There is a room here to make BIOPIC from musicians, political leaders, actors, human rights activists who had an impact on lives, from here we will have quite different interesting stories which will teach our community and can even be used for educational purposes in our schools/colleges/universities.

INNOVATION AND CHALLENGES FACING INNOVATORS IN TANZANIA

Tanzania government and its people still don't see the value of innovation and how important it is in the growth of country's economy. The innovation knowledge and the ability of thinking big outside the box and creative mind of being able to produce new technologies in every aspect of life perceptive, is

Something our government needs to focus on more especially in emphasizing this in our schools and education system. Children need to be aware and exposed to the environment that will enable them to be creative and make the most use of their mind at their very tender age. This type of environment is not there currently, so government needs to restructure its education system and also teaching skills.

This should start from having good education environment, good teachers, more practical's, and student being taught how to think and meditate and put what they are being taught at school into action.

All this cannot be done if we don't have good leaders, so for the development of our country there is need of good leaders who are willing to change and also emphasize the change on others.

Good leaders are the ones who are eager of development, the ones who understand and have a broad knowledge of things, leaders who love changes, are aware of what is happening in the world in terms of technology and who will want those technology to happen in his/her country, this kind of spirit need to come from the highest person in the country, that's a president.

We have seen USA presidents and other big nations on how they get themselves involved in their country's big projects and researches, which in future bring huge positive impact to their communities. Like President John .F Kennedy supporting NASA on their project of going to the Moon which was Successful on Apollo 11 mission.

He led first human to walk on the moon on July 20, 1969, Neil Armstrong and Buzz Aldrin landed their Lunar module (LM) that year. These types of leaders would allow changes in education system and changes of curriculum for the better generation that will bring changes. Leaders who will be the first to buy and use local products (Clothes, foods, gadget) announce publicly and be proud of that.

Changes might not be seen immediately but after sometime of investment, full support from all institutions public/private in no time huge changes of development in every area will be seen. That is what China and other Far East countries did. They invested a lot and seriously in education and technology, we are all witnesses of positive results from them. The huge advancement of technology and production of different products from China/India and South Korea (like SAMSUNG products).

For the time being this sector is kind of left back it disappoints me to see that we have number of inventors/innovators and creators from various areas that can think differently and produce a lot of amazing things but we don't utilize their capabilities efficiently.

CHALLENGES FACING INNOVATORS IN AFRICA (TANZANIA)

(a) Poor Awareness

Most of Tanzanians don't have enough awareness or they are not aware of the procedures on how to handle their Inventions/innovations/Creations in the end their work are stolen.

(b)Industry don't believe in our local inventors

There is a huge problem in most of African countries including Tanzania, in not respecting and believing in the concepts, Ideas and thought from local innovators. They Judge them way too much basing on their level of education, the way they dress up and other unnecessary aspects instead of looking at what they say or capable of providing. The society and government believes too much in white people and things coming out of Africa and ignore ours. Innovators lack total support from the government and community which made them to lose hope and quit.

(c) Some of big companies steal creative works

Another problem that I personally experienced which subsequently influenced me to come up with this book was theft of my work by a certain entity.

Instead of big companies helping young innovators and supporting in order to encourage them (like Tigo Reach For Change Campaign) , some steal their works. In some cases, a young innovator proposes an idea to a company; they in turn steal it without involving him or her in realizing it into a service or product. This demotivates innovators.

Big companies have huge role to support and encourage young innovators and recognize them to raise their spirits. If you want to sue them it becomes very difficult due to bureaucracy in legal procedures. It also needs money for advocate and other matters, thus you will find yourself wasting a lot.

PERSONAL EXPERIENCE

This book provides guideline to help young innovators to know the safe ways to follow when there is an invention, to avoid what I faced. And also to inspire African youths that we can create new things that are not in this world yet, we have that power. I believe all humans are equal and capable of bringing changes, it doesn't matter the race black or white, social differences (poor or rich) or where you are coming from? We can all do equal things if we put our mind into it, have faith on ourselves; if given enough support from both government and community all things are possible.

In 2008 I came up with a mobile service idea which was called MAJINA/SERVICE in which people were able to request by text message meaning of their names and other information and they could get that. I was in my first year in college, and I had to go to a certain XYZ company which is dealing with content service in Tanzania, they liked my Idea and they told me to bring about 1000 names (common names, Christian, Muslim) and I should do the translation to SWAHILI in the format that they want.

I did the work and I was given a one year contract which was supposed to start in May 2008, but due to technical problem, the service didn't go live.

The following year in February I resigned a contract then in March 1, my service was on for 250 per SMS. My contract had a clause which explains clearly that I was supposed to get paid monthly.

PROBLEMS

I was not paid for March, April, May, June and I was treated badly like I didn't do anything and therefore did not deserve anything. I wrote a stern letter to them threatening that I will take them to court. They responded and I was given my first payment which was as little as 20,000/= (Twenty thousand Tanzania shillings only). Without any record of how many customers used my service in March 2009. After many complaints with no action taken I took another step. I decided to go to COSOTA which is a copyright society in Tanzania, I told them the story and they advised me to copyright my work institute suit against that company.

A letter was sent to that company. They were told to meet at COSOTA where we had reconciliation meeting. The company was told to pay me the money immediately and remove all the materials I created from the system because they violated the terms of contract hence I didn't have trust in them anymore.

They paid my money but they refused to remove the content from the system. I reported the matter to COSOTA. COSOTA sent a letter to them but they refused to comply, that when they advised me to take legal action. I decided to institute civil case and build up more evidence. With enough evidence I struggled to find the advocate who will stand for my case. Since I had no money, It was very difficult for me to get one. Another challenge is that advocates in the areas of ICT are very few and hard to find.

One day I heard an announcement on the radio, that there will be an advocate exhibition at Mnazi mmoja so anyone having problem with litigation and needs assistance are invited. That Saturday morning I took a bus and went to Mnazi mmoja where I met people from TLS which are Tanganyika Law Society. I told them my problem, they understood me and that I had enough material for instituting civil case.

They advised I should go to their office few days later, which I did. They found me an advocate for free who was Dr. Paul Kihwelu, Dean of law department at Open University of Tanzania (OUT). Later I met him at ESCO LAW CHAMBER (ATC HOUSE POSTA) where we had a meeting. I told him everything so that he could build up a civil case.

In 2010 we managed to institute a civil case NO 38 at HIGH COURT OF TANZANIA, between MSAFIRI VICENT GEORGE LWIHULA Vs XYZ COMPANY in. Which I sued them for about two billion Tanzanian shillings, which remain ongoing case until now.

Status of My case at this point;

After first mediation failed, another judge (Hon. Judge Mruke) took over .The case was mentioned so many times. My Advocate Mr. Paul Kihwelu (Currently appointed as judge) was not attending for some reasons and due to work I was missing at the court too.

Finally judge dismissed the case in November 2013 due to non appearance of the parties several times. At the moment I am working on the possibilities of reinstatement of the case, because justice was not done.

LESSON LEARNT FROM THIS EXPERIENCE

- **Innovation environment in Tanzania is not supportive**

From my experience and little research I have done to other young innovators in our country, there is a lot of disappointing stories out there concerning the whole process; starting from the awareness/knowledge about the subject to the clear ways of how to follow through and also the institutions that need to support this processes are not active. COSOTA is not giving enough knowledge to the community especially to young ones from schools because they are future inventors, they only deal with adults and musicians and movie actors/actress. But they forget the roots which are children at school; because if they become aware at tender age, they will be able to do their work in very professional way.

- **Laws are not so strong and not implemented**

This has become like a national song for most local artist/innovators because for so long the art has been taken as an amateur thing in our community. The government and community have never taken it seriously, in that case even the copyright law and patent laws are not well implemented. Also art and innovation and affiliated laws have not been well addressed in the mother law (Constitution).The government need to look at this closely in order to change the way things are done.

- **Deficit of Advocate/Advocacy/Lawyers in ICT in our country**

Still on the side of law, there is a shortage of advocates in ICT matters in our country. According to my experience, it took me almost a year to find a perfect lawyer for the ICT industry; I have been searching for so long. In the 21th century where there is phenomenal advancement in science and technology, we need this type of professionals, enough and as soon as possible, to cover the needs, because problems of this nature are so common nowadays.

- **Law process is very slow**

Civil case no. 38 between Msafiri Vicent George Lwihula, and that XYZ Company was institute at high court of Tanzania in 2010, four years later the case was dismissed because of failure of my advocate and I to show up in court. But main problem here is, things were moving at a very slow pace, I suggest that Copyright and Patent issues to have their own court to make things easier and clear for the benefit of innovators.

- **It's hard to deal with big companies if you're a young innovator.**

Dealing with big companies is so difficult, because of their financial muscle. They can hire a good lawyer and pay a lot of money to play around with the evidence and at the end of the day innovator loses. So my suggestion is, for the big companies to change their perceptions towards local innovators.

They should support them, have faith in them, and help them because they are capable of changing everything and create good environment for innovators for the relationship between them is win/win, If they can create something new, it will improve company business and has to improve Innovators life too.

All of the experiences I faced in my career motivated me to write this book so that it can be useful to other young innovators. That they shouldn't learn the hard way and experience this bad situation. Instead they should enjoy the work of their innovative/intellectual abilities. A little invention might have a huge positive impact in the country in later days if it is supported and protected.

(d) Government Don't Protect/support Innovators

As I said in my foreword, we want to change and develop new Africa, new Tanzania; we want to change the quality of life of the people. Our governments has to be more serious on this matter and in order to go further it need to change, it need to legislate more and make sure they are all well implemented.

By doing this it will bring encouragement to the inventors and in few years to come we will experience the rapid changes in this sector and in our lives.

(e) Dream killers

There are people I refer them as dream killers. As Innovators or anyone who likes to grow/develop need to stay far away from them. We have these people in our families, schools, and even at work, people who think you can't do this/that and they will try their best to make sure that you don't go further with what your intuition tells you. Surround yourself with people who will encourage you, build/direct and motivate you. By doing this way you will succeed.

In Tanzania and Africa we believe in the appearance of a person and the education they have instead of what he/she can offer. So people are being judged that they can't before being given an opportunity of doing anything in the first place.

OUR LOCAL INVENTORS

In my childhood, as I was growing up in different regions in Tanzania (Morogoro, Arusha, Mbeya, Dar es salaam) I had heard a lot of fascinating stories from the street or television even radios that some of good brain Tanzanians invented good things.

I remember few of them through whom I can assert there are great thinkers in Africa/ Tanzania. People should recognize them so that we respect them and ourselves, this will help us in gaining self confidence and motivation to do more big things and hence eradicate that myth that only white people can think and innovate.

A TANZANIAN WHO INVENTED MACHINE WHICH CHANGES SMOKE TO FERTILIZER (MR. LAURENT MAKOLE),

Mr. Laurent who used to be my school mate at Pandahill Secondary school (Songwe-Mbeya), is a very talented and creative person whom I used to admire at school and even after school.He was among of leaders in our science club, and every time we met he used to show us some new things that he made and it was very inspiring.

After he left the school, I never thought we will meet again but after a year while in Dar es Salaam (Home) for holidays I saw the news on television, heard in radio and newspaper that Laurent Makole discovered a machine which can turn Smoke to fertilizer. It was a real fascinating story for me, so when I went back to school I told our academic teacher about it.

He was so excited and happy because he knew him. He told me I could have brought that news paper too but I forgot it at home. The invention was good especially in these days where there is a lot of smoke and pollution of environment (air pollution).

Later I managed to do an interview with him personally and he said he wasn't supported enough especially from government   so he didn't continue with that important technology.

NYUMBU CARS-FROM KIBAHA MILTARY BASE (AUTOMOBILE STATION)

Nyumbu car is the car designed at Tanzanian military base in Kibaha.This institution was established in 1985, Was mainly was a center owned by government to design and build African cars (Nyumbu).

They were doing a very good job when started, and we have seen number of variety Nyumbu designs from Nyumbu canters,Nyumbu boozers,Nyumbu tractor trailers and Nyumbu combine harvesters on the streets.

But nowadays we don't see them like we used to. I was expecting as time goes, more designs to come up, cars which relate to our environments for military and commercial purposes. But I don't see that kind of growth; in USA cars like Hummer were invented in the military and were used in battlefields.

But they turned them into commercial I products with a lot of designs and they are making money. I would like to see such kind of things. If this station could have been more serious, and invest more in technology, by this time we would have been producing a lot of cars in and outside our country. The station would have expanded and more Tanzanian could get employment, and for that fact our economy and life of people would have changed a lot.

[Picture of Nyumbu car]

MPEMBA EFFECT-PHYSICS THEORY (A TANZANIAN WHO DISCOVERED THE REASON WHY HOT WATER FREEZES FASTER THAN COLD WATER)

The Mpemba effect, named after Erasto Mpemba, is the observation that, in some circumstances, warmer water can freeze faster than colder water. Although there is evidence of the effect, there is disagreement on exactly what the effect is and under what circumstances it occurs. There have been reports of similar phenomena since ancient times, although with insufficient detail for the claims to be replicated. A number of possible explanations for the effect have been proposed. Further investigations will be needed to decide on a precise definition of "freezing" and control a vast number of starting parameters in order to confirm or explain the effect.

MPEMBA'S OBSERVATION

The effect is named after Tanzanian Erasto Mpemba. He described in 1963 in Form 3 at Magamba Secondary School, Tanganyika when freezing ice cream mix that was hot in cookery classes and noticing that it froze before the cold mix. After passing his O-level examinations, he became a student at Mkwawa Secondary (formerly High) School in Iringa. The headmaster invited Dr. Denis G. Osborne from the University College

in Dar Es Salaam to give a lecture on physics. After the lecture, Erasto Mpemba asked him the question "If you take two similar containers with equal volumes of water, one at 35 °C (95 °F) and the other at 100 °C (212 °F), and put them into a freezer, the one that started at 100 °C (212 °F) freezes first. Why?", only to be ridiculed by his classmates and teacher. After initial consternation, Osborne experimented on the issue back at his workplace and confirmed Mpemba's finding. They published the results together in 1969, while Mpemba was studying at the College of African Wildlife Management".

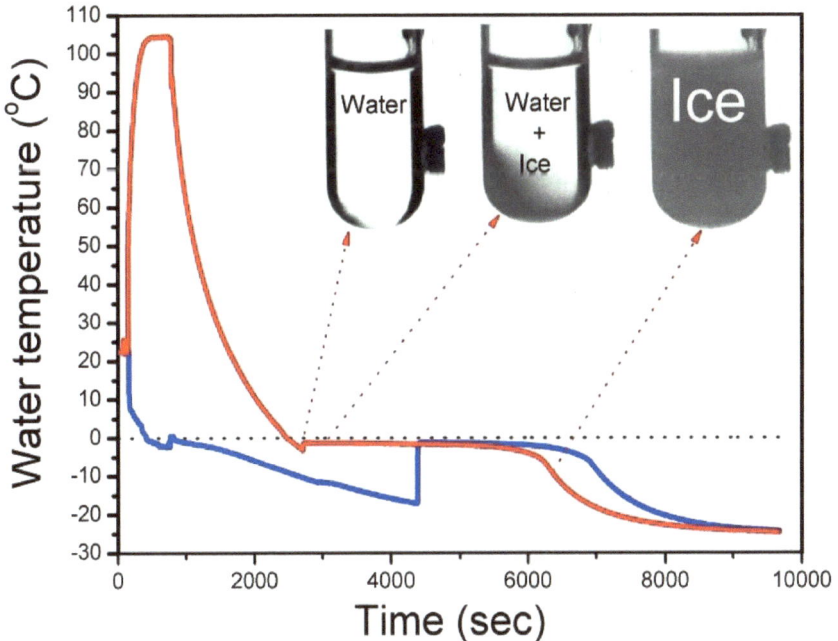

Picture of Mpemba effect graphical observation.

[Mr. Erasto Mpemba]

ASENTARABI MALIMA -A TANZANIAN WHO INVENTED A BIOSENSOR CALLED BIOLOM.

Tanzanian Dr. Asanterabi Malima who is the son of the late Professor Kigoma Malima, Dr Malima, a graduate of the Department of Electrical and Computer Engineering and now a postdoctoral researcher in the Northeastern University Center for high rate manufacturing in nanotechnology in the United States, has discovered a disease diagnostic tool that can detect diseases in their early stages.

He worked under the supervision of Prof. Ahmed Busnaina. He holds a MSc. In Mechatronics Engineering (Electronics Engineering and Computer Science). He has three patents, journals and conference publications. In 2012 together with his fellow Northeastern alumni Cihan Yilmaz, PhD of the 2013 doctoral class and Jaydev Upponi, of the previous year's class, graduates of the Department of Mechanical and Industrial Engineering, and the Department of Pharmaceutical Sciences respectively, designed a device known as Biolom.

Their device is smaller than a pinhead and has the capacity to diagnose a variety of diseases at their earliest stages. After this, they established a biotechnology firm to commercialize the device they developed at the center under the guidance of the centre's director, Prof. Ahmed Busnaina, the William Lincoln Smith Chair and professor in the College of Engineering. The biolom device consists of four distinct areas, each optimized to detect a specific biomarker indicating different types of cancer or cardiovascular disease.

The team originally developed the device to detect colorectal cancer, but it pivoted to focus on liver cancer after an exhaustive field survey of clinicians, researchers and members of the pharmaceutical industry, it said.

Think Differently Make A Difference

[Dr. Asanterabi Malima]

MSAFIRI FORMULA FOR FINDING NUMBER OF MOLES IN AN IONIC COMPOUND (ME)

I discovered a chemistry formula for finding total number of ions in a given ionic compound while I was in form three at Panda hill secondary school on 22/09/2003. I was really amazed with all this stories that I heard and some of them I have seen happening but after sometimes they fade away, there is no continuation or growing in those interesting invention; Instead of being praised and act as inspiration to our generation and future generations.

Why do they fade away? The following may be the answers;

• **The lack of government support**
It might be they wanted to go further but they needed assistance and support from government in terms of money or connections/infrastructures but they failed to get the assistance.
• **Their work have been stolen**
This is major problem in third world countries especially in our African Continent. intellectual property are not well protected, so they be easily be stolen.It could be due to lack of awareness or just some big companies steal the works purposely knowing no one can do anything.
• **They had Problems with Copyright and Patent Issues**
Another big problem innovators could face is Copyright and patent. Innovator might have done good work but if he doesn't have awareness on registering board for registering their works, and if you've e not registered, your work can easily be stolen.
• **Less support from the community**

Sometimes communities discourage innovators by not supporting their works, either by not buying their products or using their services. We see most of us African/Tanzanian people buy products made outside rather than supporting our own; this discourages local innovators/ entrepreneurs whose products fail and collapse.

- **May be they did not know what to do after they invented**

Some inventors needs directions and support in business perspectives in order to make sure their inventions bring profit to them .You might have knowledge on a certain area that made you create a certain useful product or service; but you need support from other area in mass productions and marketing also distribution and improvements.

- **They fail to turn their good inventions into innovations (Making it business).**

We are also having problems of not exposing our invention stories which is very necessary for inspiring youths. For example it's so hard to know who invented a mobile Money transfer like MPESA and what challenges he faced in doing so; this would help many youths to be inspired.

Unlike western and Asian countries, where there is continuation of their invention stories. This is simply explained by formal education we are being taught at our schools, all these are documented ideas from different people and now we go to class and get knowledge.

My point here is we need document our local inventions and put them in our school so that our children can read, understand and get inspired too.

This book is aiming at addressing the innovations challenges/problems and proposes ways to solve/reduce them, by creating enough awareness and also give directions and procedures to follow to protect your Ideas.

Because I am sure that if those Inventors mentioned above where given a certain type of support in anyway, they would have been very far by now; employing number of Tanzanian youths and reduce unemployment rate brings tax to government and our country would have been advertised internationally.

SUGGESTIONS

- **Awareness** - Should be increased especially by introducing this subject in the educational system in our country so that children should know these things at their tender age as they are growing up, and not learning in hard way like I did. For other innovators, it will help them to be strong and confident.

- Government should protect and motivate Innovators

- **Protection** - Is by legal support in creating good laws and make sure they are being implemented and in cases involving companies government should look with an extra eyes on the matters also advocate help to the innovators. Harsh punishment for those who are found guilty and a huge amount to be awarded to the beneficially

- **Motivation** - Government should motivate those who create/innovate/invent or do great things in this field, by honoring them and give them awards. This will raise their spirit and raise their game.

- **Media intervention** - Media should promote our local inventors very well as they do for our local musicians/movies stars and footballers.

Currently we have seen media in our country and other African countries supporting musician very well, which has now brought positive impact to their works and life. Now African musician are gaining respect and getting more money which is a very good thing. In Tanzania Media radios like Clouds Fm and East Africa are doing very good work in promoting our local artists.

We need to put that energy in other innovative/creative works too; we should salute and give platform to all other kind of artistic works. In a Radio/Television/news papers we need to see programs like "WE SALUTE YOU" for innovative people who do mind blowing things in the country.

We have to give them chance to be talked about in different media and also every month or every week we should salute creative people and tell the community what they did in a detailed way. I also suggest an innovative Magazine that will cover all type of innovations in our country.

We should create Awards for them, all young innovators or all innovators from all categories must be put in a competition for an award each year, a well respected award given by Government (Presidential Award).

This will raise and change everything in this industry because people will be inspired a lot and our local invention will be respected and grow fast. And the changes will be seen in a very short time. From this point you will realize how African think people will be motivated to think beyond and outside the box, and that some idea/thoughts from others will make a difference in this country.

- Easy access to data or information from government and NGOs or any other places which hold data or information which might be useful to accomplish a certain innovation. It shouldn't be an obstacle and take so long to get.

- Also for Innovators themselves they should work up and fight for their rights, they should form a union or organization that will fight for the rights of its people. Tanzanian innovators must know that even in America, people like Bill Gates faced the same challenges at some point in their career. They fought for the right of their invention not to be used freely and without recognition until they change that. And now they enjoy the work of their mind, we should do the same.

I am suggesting the following to the Tanzanian government or to the other countries which have got an embassy or high commission in our country.

I have seen, sometimes in my career that nice Ideas comes into my mind about technology or things that cannot even be done in our country due to the level of science and technology especially in computer technology, mobile phones, space technology or car model designs. My suggestion is that each embassy should have a small office which will be channeling the concept from local innovators to the specific companies or organizations in a specified country. If you have a good Idea you go there and explain to them if it makes sense the embassy should take it to the proposed company.

If it's a mobile phone Idea and you want to sell to SAMSUNG you go to the South Korean Embassy and have meeting with the proposed office if it make sense to them it will be sent direct to Samsung and if Samsung likes it, then they will come for a talk with the person who invented it; if agreed in terms that will be beneficial to both of them they do business.

This will help local innovators to be promoted internationally, also get profit and change their life and our country will benefit as well. Most of Africans have good ideas but they die with their Ideas due to lack of implementation.

The use of internet is not so safe since you never know where your idea or concept landed and you don't have anywhere to follow up because human beings are not honest.

CHAPTER TWO

MANAGEMENT OF IDEAS/DREAMS

MANAGEMENT OF IDEAS/DREAMS

"Education is not learning the facts but training our mind to think"-Albert Einstein

(a)Literature of thinking

literature simply is the written works especially those considered of superior or lasting artistic merit

PHILOSOPHY OF THINKING

Philosophy studies the fundamental nature of existence, of man, and of man's relationship to existence. In the realm of cognition, the special sciences are the trees, but philosophy is the soil which makes the forest possible." —Ayn Rand, Philosophy, Who Needs It (p. 2) or simply is the way of thinking about the world, the universe, and about society.

A philosophy is a comprehensive system of ideas about human nature and the nature of the reality we live in. It is a guide for living, because the issues it addresses are basic and pervasive, determining the course we take in life and how we treat other people.

The topics that philosophy addresses fall into several distinct fields. Among those of fundamental concern are:

- Metaphysics (the theory of reality).
- Epistemology (the theory of knowledge)
- Ethics (the theory of moral values)
- Politics (the theory of legal rights and government)
- Aesthetics (the theory of the nature of art)

The most widespread systems of ideas that offer philosophical guidance are religions such as Buddhism, Christianity, Judaism, and Islam. Religions differ from philosophies not in the subjects they address, but in the method they use to address them.

Religions have their basis in mythic stories that pre-date the discovery of explicitly rational methods of inquiry. Many religions nowadays appeal to mystical faith and revelation—modes of belief that claim validity independent of logic and the scientific method, at least for the biggest questions. But most religions are in their origins pre-rational rather than anti-rational, a story-teller's account of philosophic issues rather than a scientist's.

In Greek, "philosophy" means "love of wisdom." Philosophy is based on rational argument and appeal to facts. The history of the modern sciences begins with philosophical inquiries, and the scientific method of experimentation and proof remains an instance of the general approach that a philosopher tries to bring to a question: one that is logical and rigorous.

However, while today the sciences focus on specialized inquiries in restricted domains, the questions addressed by philosophy remain the most general and most basic, the issues that underlie the sciences and stand at the base of a worldview. Philosophy raises some of the deepest and widest questions there are. Addressing the issues in each branch of philosophy requires integrating everything one knows about reality (metaphysics) or humanity (epistemology, ethics, politics, and aesthetics).

Thinking involves the process of being aware, creative in decision making, heuristic, learning memorizing, reasoning, teaching and problem solving. Or the other way thinking is a mental process which allows beings to model the world, and so to deal with it effectively according to their goals, plans, ends and desires. Thinking involves the cerebral manipulation of information, as when we form concepts, engage in problem solving, reason and make decisions. Thinking is higher cognitive f unction and the analysis of thinking processes is part of cognitive psychology.

There are many factors that affecting thinking, which includes brain ability, gender and also environment one, is raised.

Brain ability-human being are born with different intelligent quotient depending on the genetically make up of the creature, intelligent quotient is the brain ability to reason and also understand things.

Men are from Mars Women are from Venus

Gender- It is proved by researches that on average, female brains are highly connected Across the left and right hemispheres, and connections in male brains are typically stronger between the front and back regions. Men's brains tend to perform tasks predominantly on the left-side, which is the logical /rational side of the brain. Women, on the other hand, use both sides of their brains because woman brain has a larger corpus collosum, which means women can transfer data between the right and left hemispheres faster than men. These differences make us see things in different perspectives.

There are more scientific explanations on this matter in detail which includes the following:

Processing - Male brains utilize nearly seven times more gray matter for activity while female brains utilize nearly ten times more white matter. What does this mean?

Gray matter areas of the brain are localized. They are information- and action-processing centers in specific splotches in a specific area of the brain. This can translate to a kind of tunnel vision when they are doing something. Once they are deeply engaged in a task or game, they may not demonstrate much sensitivity to other people or their surroundings.

White matter is the networking grid that connects the brain's gray matter and other processing centers with one another. This profound brain-processing difference is probably one reason you may have noticed that girls tend to more quickly transition between tasks than boys do. The gray-white matter difference may explain why, in adulthood, females are great multi-taskers, while men excel in highly task-focused projects.

Chemistry - Male and female brains process the same petrochemicals but to different degrees and through gender-specific body-brain connections. Some dominant petrochemicals are serotonin, which, among other things, helps us sit still; testosterone, our sex and aggression chemical; estrogen, a female growth and reproductive chemical; and oxytocin, a bonding-relationship chemical.

In part, because of differences in processing these chemicals, males on average tend to be less inclined to sit still for as long as females and tend to be more physically impulsive and aggressive. Additionally, males process less of the bonding chemical oxytocin than females. Overall, a major takeaway of chemistry differences is to realize that our boys at times need different strategies for stress release than our girls.

Structural Differences - A number of structural elements in the human brain differ between males and females. "Structural" refers to actual parts of the brain and the way they are built, including their size and/or mass.

Females often have a larger hippocampus, our human memory center. Females also often have a higher density of neural connections into the hippocampus. As a result, girls and women tend to input or absorb more sensorial and emotive information than males do. By "sensorial" we mean information to and from all five senses.

If you note your observations over the next months of boys and girls and women and men, you will find that females tend to sense a lot more of what is going on around them throughout the day, and they retain that sensorial information more than men.

Additionally, before boys or girls are born, their brains developed with different hemispheric divisions of labor. The right and left hemispheres of the male and female brains are not set up exactly the same way. For instance, females tend to have verbal centers on both sides of the brain, while males tend to have verbal centers on only the left hemisphere.

This is a significant difference. Girls tend to use more words when discussing or describing incidence, story, person, object, feeling, or place. Males not only have fewer verbal centers in general but also, often, have less connectivity between their word centers and their memories or feelings. When it comes to discussing feelings and emotions and senses together, girls tend to have an advantage, and they tend to have more interest in talking about these things.

BLOOD FLOW AND BRAIN ACTIVITY

While we are on the subject of emotional processing, another difference worth looking closely at is the activity difference between male and female brains. The female brain, in part thanks to far more natural blood flow throughout the brain at any given moment (more white matter processing), and because of a higher degree of blood flow in a concentration part of the brain called the cingulate gyrus, will often ruminate on and revisit emotional memories more than the male brain.

To solidify the point that we are different, Albert Einstein once said Men always are worrying about things women remember, and women always worrying about things men forget-this view from one of great philosophers in science prove that our thinking is quite different.

Males, in general, are designed a bit differently. Males tend, after reflecting more briefly on an emotive memory, to analyze it somewhat, and then move onto the next task. During this process, they may also choose to change course and do something active and unrelated to feelings rather than analyze their feelings at all. Thus, observers may mistakenly believe that boys avoid feelings in comparison to girls or move to problem-solving too quickly.

These four, natural design differences listed above are just a sample of how males and females think differently. Scientists have discovered approximately 100 gender differences in the brain, and the importance of these differences cannot be overstated. Understanding a gender difference from a neurological perspective not only opens the door to greater appreciation of the different genders, it also calls into question how we parent, educate, and support our children from a young age.

Environment - Normally if you're living around environment which people surrounding you teach how to think outside the box and environment itself forces you to solve certain problems you find yourself at that level. Also education and information helps one to connect some ideas, be able to think wide and come up with solutions easily. A thought is mental or intellectual activity involving an individual's subjective consciousness. It can refer either to the act of thinking or the resulting ideas or arrangements of ideas.

A quality thought should have the following attributes:

Accuracy

Accuracy refers to the closeness of a measured value to a standard or known value. For example, if in lab you obtain a weight measurement of 3.2 kg for a given substance, but the actual or known weight is 10 kg, then your measurement is not accurate. In this case, your measurement is not close to the known value.

Effectiveness
Is the capability of producing a desired result. When something is deemed effective, it means it has an intended or expected outcome, or produces a deep, vivid impression.

Efficiency
A level of performance that describes a process that uses the lowest amount of inputs to create the greatest amount of outputs. Efficiency relates to the use of all inputs in producing any given output, including personal time and energy.

Frugality
The quality of being economical with money or food, simply being economical with resources

Prudence
The ability to govern and discipline oneself by the use of reason simply is being cautious on what you do.

Right
Is the situation of being morally good, justified, or acceptable.

Soundness
The argument is valid (because the conclusion is true based on the premises, that is, that the conclusion follows the premises) and since the premises are in fact true, the argument is sound. The following argument is valid but not sound:

Validity
Is the quality of being logically or factually sound; soundness or cogency.

Value theory
In its broadest sense, "value theory" is a catch-all label used to encompass all branches of moral philosophy, social and political philosophy, aesthetics, and sometimes feminist philosophy and the philosophy of religion — whatever areas of philosophy are deemed to

encompass some "evaluative" aspect.

Thinking processes includes the following;

- Development of a thought
- Sharing of your thought
- Turning your thought to a product
- Challenges facing your thought/product
- Ways to improve your thinking
- The role of government and society in harmonizing all the conditions
-

Development of a thought

As explained above a thought is mental or intellectual activity involving an individual's subjective consciousness. It can refer either to the act of thinking or the resulting ideas or arrangements of ideas. All human think the quality of thinking and effects depend on our mental abilities and other external factors surrounding us, we need to train our brain to think and be able to solve problems in our areas. I believe that information, education and curious mind are keys for quality thoughts.

Sharing of your thoughts/guidelines to protect it

After coming up with an idea, sharing is necessary for clarification and proves and also how others think about it. But also we need to be careful with whom we are sharing our thoughts, because others are unfaithful, they can take them and use without your will. Due to this kind of problem it is when things like intellectual property and copyright/patent comes up. We will discuss in depth about it later in the book.

Turning your thought to a product

Thinking alone and coming up with an idea is not the end. In order for your thoughts to make an impact, one must go further than that; be able to turn your thoughts into a certain form of expression so it can be real. Expression forms can be music, concepts, book, machinery, principles or formulas. Simply is turning our unseen visions into reality so that others can understand what we painted in our mind.

Challenges facing your thought/product

There are a lot of challenges thinkers are facing especially in our African environments, which includes, intellectual property rights, society rejection, and lack of government support. But none of them should hold you back from what you believe. If you want to come up with something good and make change obstacles are always what you need to do to overcome them.

One of the respected scientists of modern time, Albert Einstein was emphasizing on being persistent and do not give up in the face of obstacles. Einstein spent years developing and testing his theories, he kept going in spite of set-backs and criticism. The following are his quotes concerning critic's challenges:

"Great spirits have always encountered violent opposition from mediocre minds,"

"I am thankfully to those entire said No, to me, It is because of them I did it myself,"
"It is not that I am so smart, It's just that stay with problems longer,"
"I think and think for months and years. Ninety-nine times the conclusions is false the hundredth time I am right."

Ways to improve your thinking

Improving of your abilities in thinking is necessary in order to faces changing challenges and problems. Thinking can be improved due to variety of things done together, includes lifestyle (exercise and relaxation), eating habit, parenting, teaching principles at school and also government support.

The role of government and society in harmonizing all the conditions

Government is necessary in providing support and making all conditions be as good as possible, like good educations, and implementations on intellectual properties laws/ copyrights and patents sometimes also recognition of the effort done by thinkers in the development of our society.

(b)Benefit of critical thinking literature review-United states of America

Hove, Genal M.

The ability to think critically is an essential life skill; current literature reveals that explicit instruction in, and practices of, critical thinking strategies in the high school classroom can improve student academic performance.

Adoption of critical thinking strategies can also prepare students for the rigors of college, as well as helping them develop the skills necessary to compete economically in a global environment.

Research on the impact of critical thinking strategy instruction in the high school English classroom supports the findings of current literature; students who receive instruction in a critical thinking strategy were better able to demonstrate critical thinking in a post-strategy instruction assessment than those students who had received no strategy instruction.

The ability to think critically is an essential life skill in American society today; as the world changes at an ever-faster pace and economies become global, young adults are entering an expanding, diverse job market.

To help young Americans compete for jobs that did not even exist a few years ago, it is necessary now more than ever before to ensure that young adults possess the thinking power to flexibly and creatively adapt to new job markets. According to Mendelman (2007), ─ the majority of

U.S. schools fail to teach critical thinking and, as a result, the majority of our populace does not practice it. Hayes and Devitt (2008) stated ─ generally, critical thinking strategies are not extensively developed or practiced during primary and secondary education. School systems need to amend curriculum to ensure that high school graduates have developed a solid foundation of critical thinking skills, enabling young adults to be more successful in their pursuits after high school.

Since the enactment of the No Child Left behind Act of 2001, pressure has been on school districts to demonstrate student progress and competency via standardized test scores. In today's accountability climate critical thinking activities can take a back seat to test preparation (Pescatore, 2007).

Rather than embarking on frustrating attempts to cram students full of simple recall facts in the weeks prior to a round of standardized tests, it may be more beneficial long-term for students to be able to utilize factual information as a framework for critical exploration of broader concepts. While it may be tempting to teach to a test, however, students don't live in a multiple choice/true or false world.

Paul and Elder (2008) insisted that multiple-choice tests are rarely useful in assessing life situation and instead teachers should develop the kinds of intellectual tasks students will perform when they apply the subject matter to professional and personal issues in the various domains of their lives. Teachers are obligated to help students develop the skills necessary to synthesize the nuances of a modern, complex society.

Beyond the personal benefits experienced by adults adept at critical thinking more

Opportunities, better jobs, higher income society also benefits when the general populace can think creatively and insightfully. According to Pescatore (2007), for social change to occur, citizens must not only think critically about what they read and view, but they must also react to transform the world. Rather than accepting information at face value, educated critical thinkers can thoughtfully explore the broader perspectives of an issue. The National Association for Media Literacy education (2010) advocated explicit teaching of critical inquiry, encouraging students in active inquiry and critical thinking about the messages that we receive and create (Oldakowski, & Sloan).

The ability of students to explore issues thoughtfully offers a way to speak out against injustice and unfairness (Pescatore, 2007). Critical thinking skills do not occur randomly or without effort; it takes structured, deliberate, and repetitive exposure and practice for students to develop insightful thinking.

Developing the ability to think critically is an essential life skill; it is also clear that practicing critical thinking strategies should be a daily occurrence in classrooms across the United States.

The high school English classroom is a logical environment in which to explicitly teach, and practice, critical thinking with the goal of developing lifelong habits of mind. As Mendelmen (2007) pointed out, If reading the world can be paralleled to reading text, then literature offers an ideal vehicle for teaching the critical skills necessary in analysis The intent of this research is to comprehensively explore current research and strategies for incorporating critical thinking into high school English curricula.

Statement of the Problem

In order for high school students to be prepared to compete for employment in a global Economy, all students must be able to think critically and strategically; unfortunately, a problem exists because critical thinking strategies are not consistently taught In American high schools, translating to a populace that is ill-equipped to easily adapt to a rapidly changing world.

Mendelman (2007) claimed that in a day and age in which more and more children grow up engaged with primarily passive activities…teaching critical reading is one of the most important and most difficult burdens of the classroom.

If students are not exposed to, and do not master, the ability to think insightfully and critically, they will be unable to compete in a modern, global economy. In order to better prepare our students for the challenges they will face, high school teachers need to explicitly teach critical thinking strategies, equipping young people with twenty-first century skills.

The high school English classroom presents a natural setting to practice critical thinking, as it is customary for English instructors to work with students on analyzing all types of text for word choice, point of view, tone, and structure to develop the skills of critical thinking — that can have clear relevance to students live (Pescatore, 2007,).

A rigorous English curriculum, focused on an explicit, scaffolded approach to teaching critical thinking skills, will better prepare high school students for college and employment.

Purpose of the Study

The objective of this study is to analyze current literature and examine strategies for developing critical thinking skills in high school-aged students. The purpose of conducting this study is to assist this researcher in implementing a structured approach to teaching critical thinking in the high school English classroom.

This research has universal implications for all subject areas, and therefore will be applicable to the general high school setting, not only the high school classroom. Mendelman (2007) asserted that critical thinking should be taught in virtually every course in the humanities.

Assumptions of the Study

Systematic implementation and practice of critical thinking strategies will help high school students develop habits of mind that allow them to view the world through a critical scope. Repeated student exposure to critical thinking practices will assist students in all academic disciplines, as well as translate to life beyond high school.

Definition of Terms

Critical Thinking: a mental process of analyzing or evaluating information, particularly. Statements or propositions that are offered as true Critical thinking can be described as a gradual progression from the superficial to the increasingly complex (Mendel man, 2000).

High-Stakes Testing: any test for which the results have serious consequences for the test taker and teacher. An example of high-stakes testing would be the Wisconsin Knowledge and Concepts Exam (WKCE); in some Wisconsin school districts, test results are being used to evaluate teacher performance.

Metacognition: the mental process of thinking about one's own thinking; the ability to assess and evaluate ones thinking. Developmentally, metacognition typically begins with the onset of adolescence.

Limitations of the Study

Potential limits to this study include the knowledge and skill of this researcher in comprehensively finding all possible research on this topic. While every effort will be made to explore this topic as thoroughly as possible, it's probable that the researcher will not examine every single bit of research on the topic. Another potential limitation is the fact that subjects may not answer questions to the best of their ability. Some parents were unwilling or unable to return the permission slip requiring their signature, which would have allowed their child to participate in this study.

Methodology

Critical thinking is an essential life skill. This paper will explore current literature on critical thinking, including critical thinking instruction and teaching strategies. To determine the usefulness of employing critical thinking strategy instruction in the classroom, research will be conducted to determine if explicit critical thinking strategy instruction results in improved performance in the high school English classroom.

Research will be conducted using sophomore English students in two classes which share the same instructor. One class will serve as the control group and will receive no strategy instruction; the target class will receive explicit instruction and practice in an differencing critical thinking strategy. A post-reading assessment quiz will be administered in an effort to determine if there is a difference in student performance.

Why Critical Thinking

For high school students to be successful in a continuously changing environment learning core subject matter is not enough; instead, core skills subject taught within a 21st century skill set is the key to student success. Students must know how to learn, how to innovate, and how to use media and technology in a career context (Pittman, 2010).

The ability to think critically is not exclusive to the academic arena; rational, reasoned thinking is an essential life skill. Critical thinking is that mode of thinking about any subject, content, or problem in which the thinker improves the quality of his or her thinking by skillfully analyzing, assessing, and reconstructing it (Paul & Elder, 2008).

Hayes and Devitt (2008) reported that the ability to demonstrate critical thinking has become so essential in today's society that it is a core competency in earning undergraduate degrees; employers of recent college graduates support this assertion, ranking strategic thinking as key factor in job success for critical thinking skills to develop, teachers need to teach critical thinking while students take responsibility for their own learning. Students need 21st century skills that allow them to own their learning; students need to be able to locate, analyze, and evaluate new information while at the same time organize and plan what to do with that new information (Coughlin, 2010, p. 50).

Critical thinking involves ways of thinking about the written and spoken word that go beyond the surface meaning in order to discern the deeper meaning, ideology, and bias expressed therein (Pescatore, 2007).

Thinking in a disciplined, critical manner does not automatically evolve on its own; educators are critical to helping students take command, and self-assess their learning and thinking (Paul & Elder, 2008). In this regard, Coughlin (2010) concluded that research on 21st century skills reveals that student success is more related to critical thinking than traditional core subject matter.

E. Benefits of Critical Thinking--Improved Classroom Performance

A classroom environment centered on a critical thinking philosophy will better prepare students for the adult world of change and uncertainty. Paul and Elder (2009) maintained that without concerted intervention and evaluation, human thinking tends to be biased, unclear and flawed. However, when we recognize this problem, this obstacle to quality in our lives, we use our thinking to improve our thinking. We use our capacity to think at a higher level to work on and improve our thinking.

Flawed thinking is then minimized. Educators using a critical thinking approach to instruction can discipline their students to continually assess the validity of their reasoning and rationale; this rigorous self-assessment best prepares students for future success.

Coughlin (2010) asserted that 21st century skills are essential qualities and will have a direct impact on student futures, including educational, professional, and life success. Ketelhut, Nelson, Clark, and Dede (2010) concluded that — curricula centered on both inquiry and coverage of state and national content standards would help teachers achieve both objectives.

Unfortunately, most students think of learning as disconnected sentences from a textbook or lecture. By the time they reach college level, they have successfully mislearned what it means to teach (Paul & Elder, 2008).

In other words, students have not developed the discipline, or the innate curiosity, to make connections between diverse disciplines. Elder and Paul (2008) are convinced that critical thinking is the key to enabling students to see the interconnected logic of any subject or specialty and to think with discipline and skill within that logic (p. 88).

According to Bernasconi (2008), high school educators adopting a critical thinking approach in their classrooms clearly appreciate the eventual demands that colleges will place on students to read and write, and most importantly, be able to think about what they read and what they write.

The irony is that some educators dismiss the teaching of critical thinking to instead focus solely on standardized test preparation, especially in this era of high-stakes testing and pay-for-performance teaching salaries.

Pescatore (2007) disagreed with this choice as critical thinking skills are useful in passing state mandated (Pittman (2010) supported this, explaining that the College Board has released detailed standards that align with expectations for entrance into core college-level course in addition to core subject content, however their standards include practical skills such as critical thinking, collaboration, problem solving and technology literacy which are key to student success in any discipline. Clearly a critical thinking approach will help students prepare for life after high school and standardized tests.

F. Benefits of Critical Thinking--Better Understanding of Self and Society

Adopting a critical thinking approach in the classroom will yield benefits well beyond academic success, especially when students are prompted to analyze their decision-making in anethical light. Pescatore (2007) advocated critical thinking instruction because it has the added benefit of fostering engagement in the public interest rather than just self-interest, enabling young people to become significant forces for change.

Without guidance and intervention, however, human beings tend to maintain narrow, self-interested perspectives (Paul & Elder, 2009). Elder and Paul (2009) feared that students receive critical thinking instruction without being challenged to clearly understand and asses their decisions in an ethical framework.

These students develop intellectual skills which enable them to get what they want without being bothered with how their behavior might affect others. By teaching critical thinking without ethics, one runs the risk of inadvertently fostering sophisticated rather than fair minded critical thinking. Critical thinking can be a powerful tool.

(c)How to manage Ideas, and guidelines

The strategically management is an interpretation of the environment, setting the company/indivual limits, defining goals and priorities and, what is most important, a role of the passion, the intuition and dreams, as they are essential like never before. The passion is the source of innovations and changes it tells to reach the future, to make ambitious objectives, cross the obvious borders break bounds of thinking.

The passion builds a motivation and a commitment which gives sense of existence, also gives the managers and workers a chance for a real commitment and a bit of a simple joy. When companies

Operate strategically, then the world of importance, dreams and hope shows in their actions beside the world of effectiveness. in academic literature front end innovation has in the last decade been given more and more attention as an area with a potential for increasing innovation capability. A line in this literature suggests exploiting this potential through the concept of idea management.

Ideas are the potential starting point for any innovation venture and by understanding and supporting idea processes in front end innovation companies can strengthen their innovative capability.

The paper aims to identify and review current literature dealing with idea management. Idea management has ancestors like the suggestion box and cousins like the ideation process but in this paper idea management will refer to the management of the process of motivating, generating, evaluating and implementing ideas on an organizational level in the context of front end innovation (literature review of idea management) Anna Rose Vagn Jensen, 2012.

The perspective is relevant because idea management is strongly related to the use of systems for capture, sharing, store and retrieving ideas, still being a complex social human process in interaction with technologies.

With this perspective view in mind the following section will review the identified literature on idea management.

In an earlier contribution on the subject of idea management, Green et al. (1983) analyses the Management of the flow of ideas in an R&D laboratory in a human information-processing perspective. Here the authors use the understanding of human information-processing as an analog, for example how the human brain processes information, synthesizes, remembers, recalls etc.

They presents a logic with human information-processing on the one side and organizational information-processes on the other side, equally contributing to the flow of ideas in industrial R&D. Managerial implications are identified, concerning the generating of ideas, capturing ideas, retaining ideas and retrieving ideas. It is interesting that this early study predicts the future of idea management and its strong connection to the use of computer technology

as an analog to the human brain. This study was before the burst of the information technology and one could only imagine how IT would take part in the work practices as idea processes and management.

The analog is interesting and when brain mechanisms are placed outside the head of people on an organizational level, interesting issues will occur in idea process practices. In a contribution at the same time, Galbraith suggests a certain design of the organization where innovation ideas, more specific radical innovation ideas, have better conditions.

The term of idea management is used on a more individual level as a Idea management literature is primarily based in cognitive and social process and concerns how ideas are developed and promoted through bargaining and negotiating in the organization.

Idea management literature is primarily based in the field of innovation management in organizations and as a part of the above described development of information technology. Idea management is also represented and developed in information technology literature dealing with applications of idea management systems. As an example, an idea management system for team creation system is developed by Xie & Zhang.

They seek to understand the process of team creation and develop a software tool to support and enhance the process. In general, the idea process of the team creation is duplicated in the tool and made manageable through the main steps of idea recognition, idea selection, idea evaluation, and idea visualization.

The work of Water ski et al. deals with the development of idea management systems and furthers it. From being nothing more than a box where employees could submit their ideas on a piece of paper, the web 2.0 techniques allow complex submission of data and data handling in idea management systems. The work of Water-ski et al suggests the use of semantic web principles to link organizational systems for better idea assessments.

The review shows that idea management knowledge is represented in innovation management literature but also in IT literature. Idea management literature primarily deals with best practice case studies and supplies a variety of frameworks, models and systems for maneuvering the stream of ideas in front end innovation.

Recent literature has begun to investigate how idea management systems are integrated in the practices of idea processes in organizations and identify certain managerial implications. There is an emphasis on both human behavior and the systems structure in managing ideas but the interplay between the two and which Managerial implications becomes relevant is still an area to be uncovered in depth.

The review leaves behind uncertainty if the idea management systems will live up to their promises of:

(c) How to manage Ideas in Tanzania environment.

If it happens a potential great/unique Idea passes through your mind and you would want it to go further, the following procedures will help you but before getting to the procedures you need to know about these things;

1. Write the idea down in a piece of paper or on your Pc for literary work
2. Create a demo or pilot for product
3. Do enough research on the matter before going to registration boards.
4. After a good research and getting, satisfied you can now go either to COSOTA or BRELA depending on the nature of the Invention.
5. After registration now you can be able to talk business with different people knowing that you have legal rights which protect your work. And if anyone misuses your work you will be able to sue him or her.

NB: If it happens that you want talk business with anyone before registration you have to use non-disclosure agreement between parties.

COSOTA-Copyright society of Tanzania which deals with copyright issues

(a)COPYRIGHT

This is exclusive and assignable legal right given to the originator for a fixed number of years to print, publish, perform, film or record literary, artistic or musical material. This is according to Tanzania copyright and neighboring act of 1999 which controls the rights of inventors through copyright society of Tanzania (COSOTA)

What do we mean by saying exclusive and assignable legal right to the originator?

The word exclusive it means having the power to limit or control or use by a single individual or group and not others without your permission. Assignable rights-are the rights that are being given to you by a legal institution which has a power to do so legally.

The Originator - in this case here we mean the inventor or the person who is the source of that new idea or concept.

So basically a copyright is a legal power given to an inventor to limit and control his work so that it cannot be misused by others without his permission. Authors of original works in which literary and artistic works shall be copyright may be entitled to copyright protection for their works under this Act, by the sole fact of the creation of such works (Copyright and neighboring rights, 1999),and copyright covers throughout owners life time and fifty(50) years after death.

Literary and artistic works shall include in particular:
- (a) Books, pamphlets and other writings, including computer programs;
- (b) Lectures, addresses, sermons and other works and other works of the same nature.
- (c) Dramatic and dramatic-musical works
- (d) Musical works (Vocal and instrumental). Whether or not they include accompanying words;
- (e) Choreographic works and pantomimes;
- (f) Cinematographer, works, and other audio-visual works;
- (g) Works of drawing, painting, architecture, sculpture, engraving, lithography and tapestry.
- (h) Photographic works including works expressed by processes analogous to photography;
- (i) Works of applied art whether handcraft or produced on an industrial scale
- (j)Illustrations, maps, plans, sketches and three relative to geography; topography, architecture or science.

All works should be protected irrespective of their form of expression, their quality and the purpose for which they were created.

The following shall be derivative works protected as original works;
- (a)Translation, adaptations, arrangements and other transformation of the literary and artistic works.
- (b)Collection of literary and artistic works, such as encyclopedia and anthologies, or collection of expressions of folklore and compilation of data or data bases which, by reason of selection and arrangement of their contents constitute intellectual creation.

Protected protection shall not extend to
- (a)Laws and decisions of courts and administrative bodies as well as to official translations thereof;
- (b)News of the day published, broadcast or publicly communicated by any other means.
- (c)Any idea, procedure, method of operation, concepts, principle, discovery or mere data, even if expressed, described, explained illustrated or embodied in a work.

The author shall have the exclusive right to carry out or to authorize

The following acts in relation to the work:
- (a) Reproduction of the work
- (b) Distribution of the work
- (c) The rental of the original or a copy of an audio-visual work, a work embodied in sound recording, a computer program, a database, or a musical work in the form of notation, irrespective of the ownership of the original or copy concerned.
- (d) Public exhibition of the work
- (e) Translation of the work
- (d) Adaption of the work
- (e) Public performance of the work
- (h) Broadcasting of the work;
- (i)Other communication to the public of the works
- (j) Importation of copies of the work (from Tanzanian copyright and neighboring rights, 1999)

Filling of a copyright

If you have your work and you need to file it for copyright, you go to copyright society of Tanzania which is located at Mikocheni, Dar es Salaam. And register for membership first, there is a little fee which is manageable, after that you can be able to file your works so as the office can evaluate and research on your invention and see if you deserve to get copyright.

For an inventor to profit for his work and to earn respect and recognition copyright is very important, let's see the case study of how copyright helped Bill gates to succeed.

Why is Bill Gates So Rich

Bill Gates probably has an above-average level of intelligence. He definitely had a good idea. It helped that both his parents were lawyers—at least he knew that he could trust his attorneys and that they were arguably able to find the type of specialized legal counsel he needed.

This helped ensure that his contracts were enforceable and that these contracts considered the most significant issues respecting his rights. However, Bill Gates would be, at best, a freelance computer programmer today, were it not for the Copyright Laws.

When I say the Copyright Laws, I don't just mean those of the United States. I include those of all industrialized nations and the corresponding infrastructure which helps ensure that a person's copyrights are respected. These laws, though regional, basically protect the author or computer programmer to a similar extent.

Blatant copying of a copyrighted work is protected. Penalties for copying are often very stiff. For example, in the United States, recoveries range from statutory damages from between $20,000 and $100,000 (if the copying was intentional and the author registered his copyright with the Copyright Office) to an unlimited amount, based on actual damages.

Why is the Copyright Laws responsible for Bill Gates's wealth? Because the underlying program code of DOS was protected from copying under the Copyright Law and, later, DOS became the standard operating system for the personal computer industry. The original code was the blue print of codes to come, making just about any duplication of the underlying code (the duplication being almost mandatory in the development of DOS compatible software) an infringement unless Microsoft's Copyrights are respected, i.e., through entering into a licensing agreement with Microsoft. Such licensing agreements included clauses which prohibited the reverse engineering of the code, thus further protecting Bill Gate's intellectual property under the Contract Law.

On top of all this, a copyright is valid for the life of the author plus 50 years or, as in this case, either 75 years from publication or 100 years from creation, whichever expires sooner (due to the fact that the code as it was developed, was a "work for hire", i.e., written by employees of Microsoft). This validity term is much longer than the Patent Laws provide protection (maximum of 20 years from a patent application's filing date).

This means that it will be the middle of the next century before anyone can identically copy the original DOS code. And it will be 2073 before anyone can copy Windows 98™. In other words, the value of the code will have long expired before the protection runs out. This, combined with the overlapping, interdependent nature of upgrades, results in a perpetual monopoly.

Some might rightly argue that other factors had a strong influence on the explosive growth of the Gates fortune. However, legal experts have considered the fact that applying the Copyright Laws to an industry which is epitomized by technology may not be appropriate. After all, software is much more like hardware than it is like a novel, a song or a play.

Experts and other interested parties have recognized this fact and successfully lobbied the United States Patent and Trademark Office and European Patent Office to permit the issuance of patents for computer software-related inventions. However, the protection offered by Copyright is concurrently available and often more valuable than that offered under the Patent Laws.

The Patent Laws, however, promise broader, complementary protection for the programmer. Unlike the Copyright Laws, the Patent Laws protect against use of the same method as that of the owner or exclusive licensee of the patent.

Conclusion: Consult with an Intellectual Property Attorney early and consider all the tools you may have at your disposal to best ensure that your high tech startup will succeed.

IN DETAIL ON THINGS THAT CAN BE COPYRIGHTED:

(1)MUSIC

-Music is the vocal or instrumental sounds (or both) combined in such a way as to produce beauty of a form, harmony, and expression of emotion.

In music we have a good example whereby in 2012 there was an American/Hollywood film released (PEEPLES) in which they used a sound track of our own Tanzanian traditional song musicians SAIDA KAROLI massive hit MARIA SALOME, in this case they should have been prior agreement between the artist and those responsible for the movies otherwise SAIDA KAROLI can file for the copyright case.

2) MUSIC NOTATION

Music notation or musical notation is a way of writing down music so that anyone can play it. Many systems have been used in the past to write music. Today most musicians in the world write musical notes on a stage,, five parallel lines with four spaces in between them,(simple English Wikipedia)

PICTURE- MUSIC NOTATION

(3)LYRICS

Lyrics-are the set of the words that make up a song usually consists of verses and choruses.

KAMA HUWEZI LYRICS-RAMA DEE FEAT LADY JAY DEE
Verse 1
LedyJadee
Vipi mpenzi huwoni mbali
Kweli huwezi kwenda mbali na miee
Kweli mengi huletwa ndani
Si halali kuyapokea yote unitupie
Tutagombana kilasiku mpenzi
Tutaudhiana kilasiku ndani
Tutatishana kugawana mali
Bila kumjuwa mchawi wetu nani

Music industry in Tanzania is growing fast employing a lot of youths, life has helped to change their lives and also government get revenues through their works, also help in the growth of language and popularity of our country, outside the boundaries currently Diamond platmumz won 3 channel O awards in south Africa,

[Picture of Diamond Platinumz after receiving three Chanel O Awards]

(4)INSTRUMENTAL

Instrumental is a piece of music performed by instruments with no vocals. Or it's a musical composition or recording without lyrics in it, it's sometime called a Beat.

(5)MOVIE

Movie or film or a motion pictures is a series of still images which when shown on a screen, creates illusions of moving images.

Example:
-Moses – Steven Kanumba
-Dj Ben- Steven Jacob (JB)
-Fake Pastor- Ray kigosi

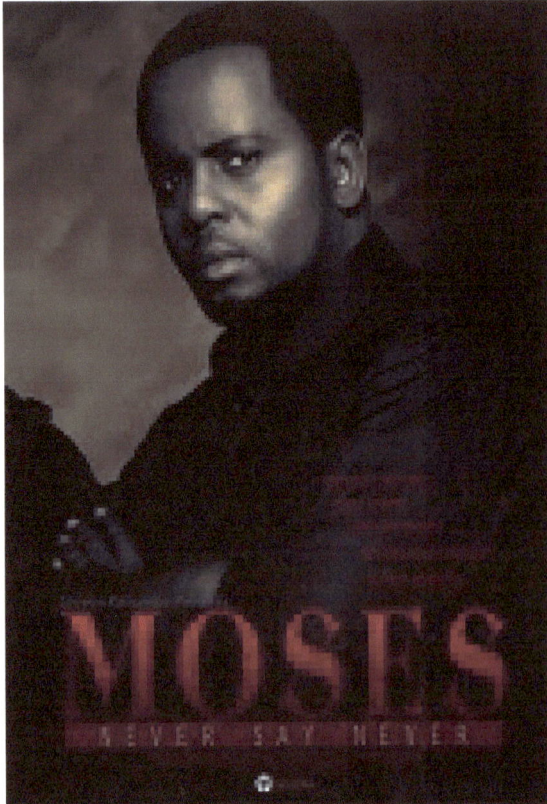

(6)MOVIE SCRIPT

Movie Script-it's a Screen play or Script is a written work by screen writers for a film, video game, or television program. These screen plays can be original works or adaptations from existing pieces of writing. In them, the movement, actions, expressions, and dialogues of the characters are also narrated.

INT. DEREK'S HOUSE

Derek talks on the phone.

>   DEREK
>   All right, I'll talk to you tomorrow then.
>   Bye.

He hangs up. From the other room, we hear Derek's step dad, Jeff, yell. Jeff has a distinctive Texas accent.

>   JEFF
>   Derek! Derek take out the trash!

Derek mocks Jeff's accent.

>   DEREK
>   Derek, take out the trash. Take out the
>   trash Derek.

Derek comes to the realization that this is the accent he's been looking for.

>   DEREK (CONT'D)
>   (Stands up)
>   I am Major General James Stratford and
>   you WILL take out the trash!

>   JEFF
>   What?!

>   CUT TO:

FILMING DAY TWO: VIETNAM

IN CAMERA

The boys are in the Back 40. John and Harrison are dressed as soldiers, playing cards, and Derek is a commander. Will is in his uniform with a duffel bag.

>   DEREK
>   Welcome to Vietnam, boy! I'm Major
>   General Jim Stratford, but everyone calls
>   me Bowie.

>   (CONTINUED)

[Example Movie script]

(7)DATABASE

Database- Is a structured set of data held in a computer, especially one that is accessible in various ways. It can be a well programmed spread sheet, MS access small database, MySql Data base or Oracle database.

(8)LOGO/TRADEMARK

Is a symbol or other small design adopted by an organization to identify its products, uniform or vehicle.

[Picture Examples of a Logo.]

9)LITERARY WORKS

A literal work is- It is an intellectual work expressed in written words, numbers or symbols (but not audio-visually) in any medium.

Example.
Plays
Which is a play is form of literature written by a playwright, usually consisting of scripted dialogue between characters, intended for theatrical performance rather than just reading (en.wikipedia.org)

Dramas
Which is the specific mode of fiction represented in performance,(en.wikipedia.org)
Screen plays.
Which is written work by screenwriters for a film, video game, or television program, these screenplays can be original works or adaptations from existing pieces of writing, in

movement, actions, expressions and dialogues of the characters (en.wikipedia.org)
(10)BOOKS
A book- is a set of written , printed, illustrated or a blank sheet made of ink ,paper, parchment, or other materials usually fastened together to hinge at one side.
(11)WEBSITE
A website is also written web site or simply a site is a set of related web pages typically served from a single web domain.

Example.
www.jamiiforum.co.tz
www.vitabustore.com

(12)SOFTWARES
Software is an entire set of programs, procedures, and routines associated with operation of a computer system, the term differentiates these features' from hardware's, the physical part of computer.

Example of software.
-Application software's-Like Microsoft office.
-System software- Operating system like Windows, 95, 98, Xp, Vista, 7, 8.
-M-PESA Systems
-ARPsystems like, Oracle, SAP
(13)PAINTS&DRAWINGS
A Paint is a colored substance which is spread over a surface and dries to leave a thin decorative and protective coating.

[Picture Example of paint.]

A Drawing is a form of a visual art that make use of any number of drawings instruments to mark a two dimensional medium. Instrument used include graphite.

[Example of Drawing.   ]

(14) POEMS
Is a piece of writing in which expression of feelings and ideas is given intensity by particular attention to diction (sometimes involving rhyme) rhythm, imagery.
Or
Poetry

Poetry (ancient Greek: ποιεω $(poieo)$ = I create) is an art form in which human language is used for its aesthetic qualities in addition to, or instead of, its notional and semantic content. It consists largely of oral or literary works in which language is used in a manner that is felt by its user and audience to differ from ordinary prose.
Or

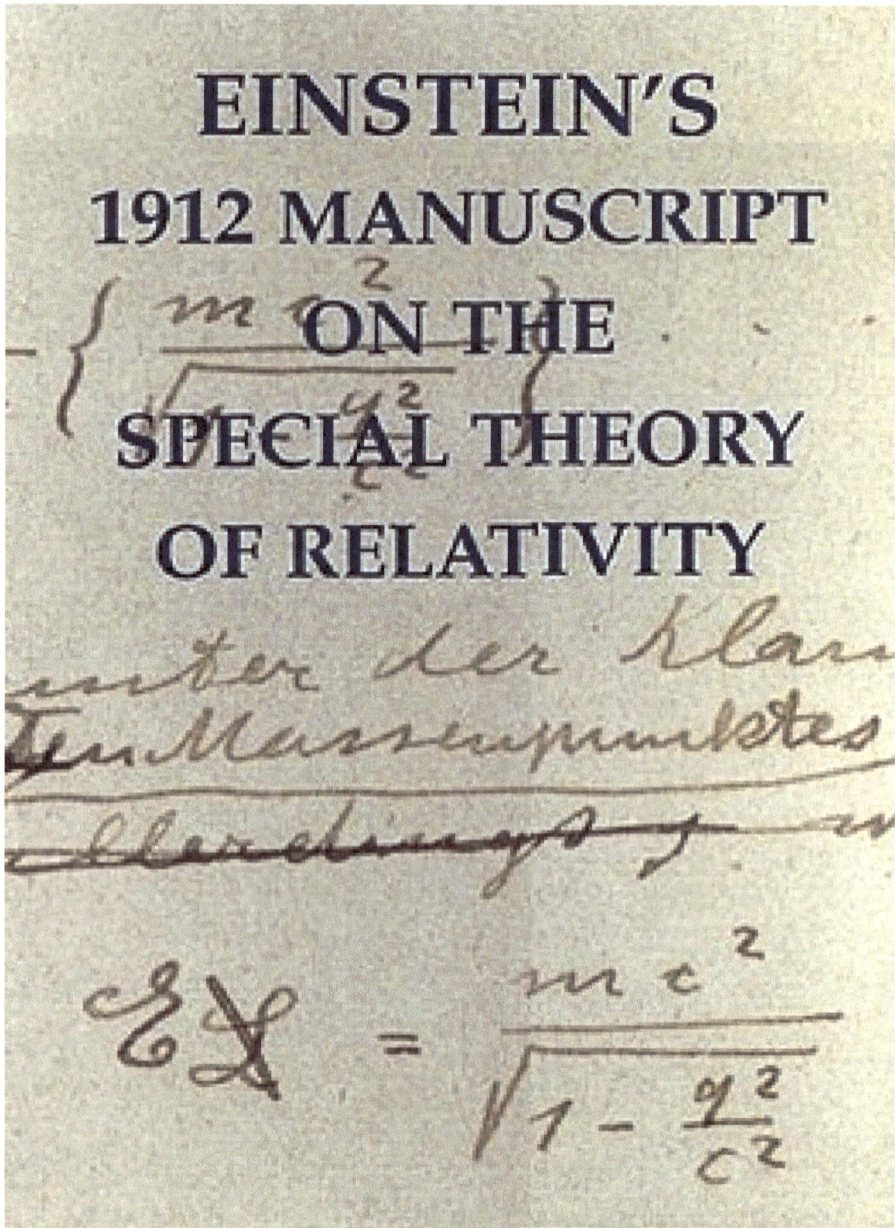

EINSTEIN'S
1912 MANUSCRIPT
ON THE
SPECIAL THEORY
OF RELATIVITY

1. A verbal composition designed to convey experiences, ideas, or emotions in a vivid and imaginative way, characterized by the use of language chosen for its sound and suggestive power and by the use of literary techniques such as meter, metaphor, and rhyme.
2. A composition in verse rather than in prose.
3. A literary composition written with an intensity or beauty of language more characteristic of poetry than of prose.
4. A creation, object, or experience having beauty suggestive of poetry.

(15)CONCEPTS
A concept is a general idea derived or inferred from the specific instances or occurrences. Something formed in the mind a thought or notion.

(16)MANUSCRIPT
A manuscript is a book, document or piece of music written by hand rather than typed or printed.

COSOTA Contacts in case for anything;
P.O.Box 6388
Chato Street
Regent estate
Dar es salaam-Tanzania
E-mail: ceo@cosota-tz.org
Website: www.cosota-tz.org
Landline: +255 22 2700019

(b) PATENT

This is a government authority or license conferring a right or title for a set period, especially the sole right to exclude others from making, using or selling Invention.

The word license means is a permit from an authority to own or use something, to do a particular thing or carry on a trade.

So in here it means that government is giving an authority to an inventor to own the invention in business issues in a certain period of time and restricting others to use it. This is according to Tanzanian patent acts no. 1 of 1987.

ACCORDING TO WORLD INTELLECTUAL PROPERTY ORGANIZATION (WIPO)

A patent is an exclusive right granted for an invention, which is a product or a process that provides a new way of doing something or offers a new technical solution to a problem.

It provides protection of the invention to the owner of the patent, the protection is granted for a limited period generally 20 years.

Types of protections which patent offers include the following;

(a)An invention cannot commercially be made, used, distributed or sold without patent owner consent.

(b)Patent rights can be enforced to the work which is a system that is capable of stopping patent infringement.

Patents provide incentives to individuals by offering them recognition for their creativity and material reward for their marketable inventions. These incentives encourage innovation, which assures that the quality of human life is continuously enhanced.

A patent owner has the right to decide who may or may not –use the patented invention for the period in which the invention is protected. The patent owner may give permission to, or license, other parties to use the invention on mutually agreed terms.

The owner may also sell the right to the invention to someone else, who will then become the new owner of the patent. Once a patent expires, the protection ends, and an invention enters the public domain, that is, the owner no longer holds exclusive rights to the invention, which becomes available to commercial exploitation by others.

The African Intellectual Property Organization

Under such regional systems, an applicant requests protection for the invention in one or more Countries, and each country decides as to whether to offer patent protection within its borders. The WIPO-administered Patent Cooperation Treaty (PCT) provides for the filing of a single international patent application which has the same effects as national applications filed in the designated countries. An applicant seeking protection may file one application and request protection in as many signatory states as needed.

Role of patent in our daily lives:

Patented inventions have in fact affected every angle of human life, from electric lighting (Patents held by Edison and Swan) and plastic (Patents held by Baekeland) to ballpoint pens (patents held by Biro) and microprocessors (Patents held by Intel)

All patent owners are obliged, in return for the patent protection, to publicly disclose information on their invention in order to enrich the total body of technical knowledge in the world. Such an ever-increasing body of public knowledge promotes further creativity and innovation in others. In this way, patent provides not only protection for the owner but also valuable information and inspiration for the future generations of researchers and inventors.

How patent is filled?

The first step in securing a patent is the filling of a patent application. The patent application generally contains the title of the invention, as well as an indication of its technical field; it must include the background and a description of the invention.

To name few patents can be for:

(1)MACHINES

Machines are a tool containing one or more parts that uses energy to perform an intended action. Machines are powered by Mechanical.

Examples a machine (Masthead-design-rotating-machines-masthead

(2)ENGINES

An engine or motor is a machine designed to convert energy into useful mechanical motion. Heat engines including internal combustion engines and external.

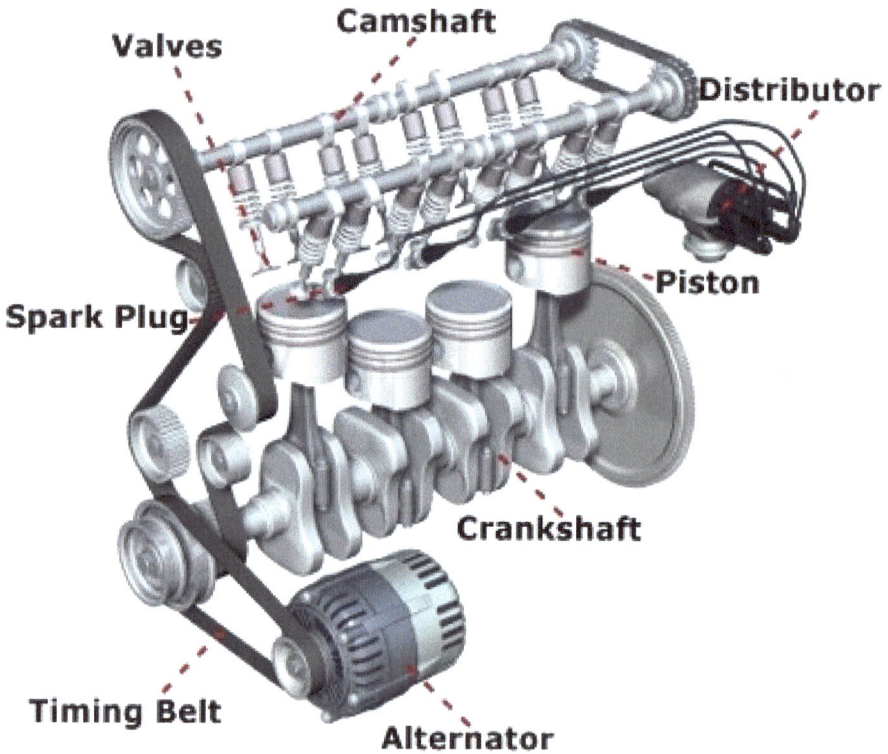

[Picture of A car engine]

So if it happens you invented your own type of engine it can be patented and you can have the patent of your own.

(3)CHEMICAL FORMULAS-
Chemical formula is a way of expressing of information about proportions of atoms that constitute a particular chemical compound by using a single line.
Includes:
(a)Medicinal formulas
(b)Detergent formulas
(c) Herbal formula's
(d)Cosmetics formulas
(e)Food formulas.

BRELA Contacts in case for anything;
P.O.Box 9393
Ushirika(TFC) Building
Lumumba street
Dar es salaam-Tanzania
E-mail: ceo@brela-tz.org
Website: www.brela-tz.com
Landline: +255 22 2180411

Academic Inventions
-Once someone's invented any new academic materials either a formula/principle or anything relating to academic you should try to communicate with COSTECH (Commission of science and Technology or University of Dar Es salaam, in a specific department of an invention) so as to publish the invention on the International journal and to see if the invention is really New that as you can be recognized locally and internationally by academicians. Contacts for the commission of science and technology are:

COSTECH-(Tanzania Commission for Science and Technology)
P.O.Box 4303
Ali Hassan Mwinyi Road
Kijitonyama(Sayansi) Costech Building.
Dar es salaam, Tanzania
Email:info@costech.or.tz
Website: www.costech.or.tz
Landline: +255 22 2927538

CHAPTER THREE

INNOVATIVE IDEAS TO IMPROVE
PERFOMANCE&CREATIVITY AT SCHOOL

INNOVATIVE IDEAS TO IMPROVE PERFOMANCE&CREATIVITY AT SCHOOL

(A) LET YOUR CHILD BE DIFFERENT

From babies' class, kindergarten to primary school, parents must know their children how they spend their day, i.e. what are they doing? Which kind of activities they are being involved? Are those activities good or bad to them, are they developing them or destroying them. Good children don't just happen by accident, some procedures should be followed, and it is parent obligation to shape his /her child before even teacher does that.

Parents should keep an eye closely to their kids and direct them what to do, also be able to know what are they taught at school and see if your child is moving towards the right direction. Be close to his/her teacher so that you can know daily progress of his or her and discover his/her strength or weakness.

TEACHING PRINCIPLES TO RAISE INNOVATIVE ABILITY

(B) TEACHING PRINCIPLES.

In Tanzania Many teachers still assume that creativity is innate and random. They believe that some people are struck by it, or maybe they have a special gene. Teachers are seldom taught how to teach creativity, but a few teachers have a 'gift' for it.

Before the industrial revolution, work and creativity was woven into the fabric of everyday life. During the industrial age, only a few elite innovators were needed to keep things going. Most jobs just required routine tasks from dawn to dusk. Middle management could learn standard procedures in college.

Universal public education was developed to prepare workers with basic knowledge and some making skills. In post-industrial times, work is again being woven into everyday life. Creativity plays a larger role than ever before for the average person, but schools are teaching less creativity.

Will universal public schools fade away like shop classes at the end of industrial life, or will they transform themselves into a new life form to meet new needs? Through all this, education in the visual arts has developed and retained a subset of educators who have gone against the dominant current of predetermined concrete knowledge and skill standards. Art is a process. It is a search. Art depends more on questions than on answers.

Therefore, we have paid attention to learning that molds the mind's thinking habits in the direction of creativity. Teachers should leave the old way of teaching student in order to pass exams; instead they should develop their understanding and critical/analytical thinking.

I see most teachers are using this method, this method make student memorize what they have been taught at school in order to pass exams and they don't understand the topics.

Disadvantages

- It kills creativity ability
- It makes people have a rigid mind and not flexible
- It makes students not to reason in solving simple matter.

I suggest that for a teacher of any subject he/she should make a thorough reading on the topic before standing in front of student and feeding them.

Teacher should at least know what his or her teaching "THE MATTER" history; its founders, its origin, its challenges to where it is now. This will make the student interested with the topic, also raise their spirit and motivate them; because some of the very great invention in life comes from just normal people.

Tell them that everybody can invent and everybody is genius, Albert Einstein once said everybody is genius in his/her way because if you judge the ability of a fish climbing tree it will live many years believing it is stupid.

EVERYBODY IS A **GENIUS.** BUT IF YOU JUDGE A FISH BY ITS ABILITY TO CLIMB A TREE, IT WILL LIVE ITS WHOLE **LIFE** BELIEVING THAT IT IS **STUPID.**

— ALBERT EINSTEIN

Think Differently Make A Difference

A scientist called Michael Faraday (book binder) was just normal person with even less knowledge on electricity and he was a clerk in the laboratory but one day he recommended something remarkable due to interest he has on the matter.

The point is everybody can invent, everybody can innovate and everybody can create we shouldn't undermine the brain of any human being; all we need is inspiration and examples and supportive environment.

These are the ways our colleagues in western countries are doing; they support things like this, it doesn't matter how small it is? If it has a positive effect they push it and make it grow big. A lot of examples in this context. We need to change our perception and priorities in the things that seriously need to be focused with high lens from both government and the community.

At schools student should be involved in much discussions between themselves and teachers about different topics of interest; and there should be no fear or discouragement from teachers and other student. Children should be motivated in anything they seem to be interested in which has positive impact.

(b) Sports and Exercise/involving in social activities.

Students need to be given enough time to play different games they love, because exercises have scientifically proven to help improve health and activate different parts of the body.

When you feel good you do well, because your mind works properly and this will help kids to understand what are being taught in class and analyze them.

(c) Putting students at tension

Putting student under a certain tension for the purpose will help you determine their thinking level, and what they can do to resolve problems.

Ask them logic questions, sometimes questions that are out of the syllabus topics so as to see how they respond to it; these will help them to a wide range of studying so as to know many things.

Exams questions should not be very straight, questions should be a bit tricky and which requires connecting some knowledge in order to find the answer.

Teachers should be very friendly to students; this will create less fear between them. For example in my case, the chemistry teacher at my school was very friendly and encouraging. I wish every student would get those types of teachers for every subject.

He will come in the class and tell us interesting science stories, he told us we can contribute anything in the discussion; he valued our contribution and it was good, we got motivated every time he came in the class. Teachers and schools should honor students who are creative and who show potential to become better one. Nothing pushes students than honoring and respecting the little contribution they bring in.

(d) Spiritual knowledge

I believe that our almighty God is the number one creator, inventor and innovator, you can see how the world and universe and everything in it was made,(Read bible-Genesis). So he passed on that creativity to us, he made us able to create. Here is my point, if we are capable of creating, we should be taught morals of creation, because without human faith/morals people can create monsters and devils.

Not all inventions are good inventions. Let's look the issues of human cloning; which is creating an exact copy of human being by the use of a part of the cell of human being. This was never acceptable in our society because it is against our faith and morals as human being.

Cloning would have created problems of using other humans as toys or machines, and other outcome that we do not yet know. Because this cloned human will lack spiritual assistance to know whether this action is good or bad in terms of spiritual reasoning and humanity. So as we teach our children/youths the whole concept of creativity we must induce them with morals /ethics of the subject too.

(e) Academic knowledge

Academic knowledge is important because that's where a child understands. Make sure children are supported with good environment which help them to study, schools, library/ books.

Because developing your reasoning abilities comes with information, which we can get through schools, reading books, listening to people/radio, watching television and also personal experiences. Parents should provide children with educative books, build the tradition of reading in them, at their tender ages try to read to them some fairly tales books, so that build the sense of imaginations in them. It's through imaginations when people usually connect the dots and come up with solutions.

(f) Biological understanding

From biological point of view as I said in earlier chapters, that male and female human beings differ, let them understand who they really are? And know their potentials and focus on their potentials in order to tap their abilities and utilize them from early age.

Parents have a task of trying to figure out at early age of their children to discover their interest and abilities in order to support them.

CHAPTER FOUR

ROLES IN FOSTERING CREATIVITY AND INNOVATION

ROLES IN FOSTERING CREATIVITY AND INNOVATION

(a) Parent/guardian role

The whole concept of reproduction and conception of a new human being is a bit complex with an amazing biological knowledge behind it. Basically the new being is a product of genetic make up of two beings (Male and Female). The character and behavior of a new born baby totally depends on the genes or genetic makeup of his or her parents.

Also there are other external and internal factors that might affect child abilities in different ways which I am trying to look at in this chapter so as to avoid some defects on the unborn child. Thus he/she can be as good as we expect him/her to be in the future and also be useful to the society.

Pregnancy period is critical and need to be watched carefully with an extra eye first knowledge is needed about the baby within a mother womb, so that mother protects her baby. She must be aware of the things that can be good or bad to her child. So that she do good and avoid the bad ones.

BASIC POINTS TO WATCH OUT DURING PREGNANCY.

{A} Watch out what you eat (Avoid junk food)

Babies are prepared from the time reproduction process starts. On the other hand it is from when sperm meets an egg for fertilization. From there onwards as a woman, you need to watch the type of food/drinks you take in and things you do(Lifestyle).

{B} Watch out Medicine you take during pregnancy

Also watch out for the type of medicine you use when you fall sick. Some medicine or strong medicine has harmful effects on the baby.

Because everything that you do at this vital time in the process of making of your child has a direct impact or indirect effect to the baby born later.

(C) Involve yourself in exercise

As pregnant woman, try to involve yourself in exercise and sports which are advisable by doctors.

(C) Avoid taking alcohols and cigarettes

-Because at this stage mother and child are sharing a lot of things, the foods she eats or things she drinks and smokes. They are absorbed into the blood stream and get to the baby, and since the baby has no immunity and everything is at building up stage; will affect direct. Also effect might be seen later in either deformation of some body parts and also ability of brain which is my exact concern, because this is where intelligence is.

(d) Stay away from too much cosmetics

Avoid using many cosmetics and sprays because some of them contain harmful chemicals which might affect the child you're carrying.

Don't be too idle or lazy, exercises helps a lot in the body in terms keeping body active, the flow of blood, avoid disease attacks that may affect you and your child. And also prepares a mother for the safe delivery of health baby; so it's crucial at this moment of life.

Proposed exercises;

*   Running short distance
*   Walking normally in the morning and evening.

Try to be alert for any signs, changes or symptoms that might occur during this period, because anything might have direct impact to your baby. You will be getting other directions and assistance as you visit doctors/clinics. As a pregnant woman you should be curious about anything concerning pregnancy from other people who have passed through that period; and try to read books so that you discover stages and things you are supposed to know/do at a particular time.

(b) WAYS TO BOOST YOUR CHILDREN BRAIN ABILITY

In this chapter we will be focusing on the early pregnancy stage and health of the baby after she/he is born especially on the area of food, nutrition and medication. This is very crucial stage of the baby's growth because it's in this stage when brain and every important part of the body which is responsible with intelligence is developed.

In their tender ages children are advised to live a certain kind of life style including eating certain kind of food because "YOU ARE WHAT YOU EAT". So if parents are not careful in watching what their children are eating; they are preparing for a bomb which explodes.

Sometimes in your pregnancy you may be feeling sick and you will need to use some medications. Some are safe but others are not, their effects cause problems and affect the child brain.

Medicines which are safe to be taken during pregnancy include;

Prenatal Vitamins are safe and important to take when you're pregnant, ask your health care expert about the safety of taking other vitamins, herbal remedies and supplements. Most herbal preparations and supplements have not been proven to be safe during pregnancy.

Another important thing to watch and be serious about for newborn baby is lactation for growth of health brain and immunity.

LACTATION

Lactation is medical term for yielding of milk by mammary gland which leads to the breast feeding. Human milk contains ideal amount of nutrients for the infant and provides important protection for the diseases through the mother's natural defense.

Breast milk cannot be duplicated by commercial baby food formulas, although both contain protein, fat, and carbohydrates. In particular breast milk changes to meet the specific needs of a baby.

The composition of breast milk changes as the baby grows to meet the baby's changing needs. Most important, breast milk contains substances called antibodies from the mother that can protect the child against illness and allergies Antibodies are part of the body's natural defense system against infections and other agents that can cause disease. Breast milk also helps a baby's own immune system mature faster. As a result, breast-fed babies have fewer diarrhea, fewer ear infections, rashes, allergies, and other medical problems than bottle-fed babies.

There are many other benefits of breast milk. Because it is easily digested, babies do not get constipated Breastfed babies may have fewer speech impediments, and breastfeeding can improve cheekbone development and jaw alignment.

In addition, breastfeeding does not involve any formulas, bottles and nipples, or sterilizing equipment. Breast milk is free, and saves money by eliminating the need to buy formula, bottles, and nipples. Because breast-fed babies are healthier, health care costs for breast-fed infants are lower.

A nursingwoman should always check with her doctor before taking any medications, including over-the-counter drugs.

These drugs are not safe to take while nursing:

- Radioactive drugs for some diagnostic tests
- Chemotherapy drugs for cancer
- Bromocriptine
- Ergotamine
- Lithium
- Methotrexate
- Street drugs (including marijuana,heroin,amphetamines)
- Tobacco

Understand your child whether is Introvert or Extrovert?
This will help you to raise and control your child-article from Parent Society by Michelle Seize of 21/October 2014, says

I've been an introvert since the beginning, way before I knew what the word meant. As a child, I thrived on being alone and spent many hours in the playroom pretending to be a teacher, or in my bedroom reading books and drawing pictures. As an adult, I do enjoy spending time with other people, but I still find that being alone is my best recharge.

How you recharge is essentially what separates the extroverts from the introverts. Introverts tend to be shy, but they don't have to be. Extroverts tend to be outgoing, but that doesn't always mean they have thousands of friends. Where you get your energy, your spark, is what really sets a person apart as introverted and extroverted. If being around other people fills you up, you're most likely an extrovert. If after spending time with people you need time alone to refill, you're an introvert.

So think about your kid(s). Whether they have siblings or are the only child, do you get the sense that being around people is fun for them, but also exhausting? Do they retreat to their rooms often — and not with friends, but to be alone?

Do they seem more vibrant and happy after they've had a weekend of play dates, soccer games, and other social events? When they've had a good week at school with their friends, do they seem more motivated to do their chores and homework?

Knowing where your kids stand is important, both in knowing how to parent them (a consequence for bad behavior of no phone time or friend time would probably be a better motivator for an extrovert than an introvert) and in teaching them self-awareness. And the earlier they learn about their personal boundaries, their needs, and what makes them unique, the better.

Teaching these concepts from a young age, and equipping our children to embrace who they are be it an extrovert or an introvert hopefully shapes a confident pre-teen, teen and eventually young adult who has a healthy self-esteem and is capable of engaging in positive relationships and activities that they enjoy, not that they feel they must do simply because of a friend or parent's influence.

This self-awareness can also pave the way for your child to have a more successful academic career. If your child is an introvert, could she benefit from one-on-one tutoring, Cyber School or other types of independent study courses? If your child is an extrovert, would a homework buddy/group motivate him to learn?

Taking these steps now may also shape their future career choices. It's never too soon to encourage our children to pursue careers that fit their personality best, jobs that will bring them fulfillment and give their life purpose… not just the ones with the biggest paychecks or the most bling.

I'm not sure where my 4-year-old stands yet. She clearly enjoys being around adults and kids and loves to meet new people. But there have been times when I've seen her retreat in social situations, and times when she seems to express the need for a "stay at home and do my own thing" kind of day. Time will tell, and I look forward to helping her find the right balance either way.

Also it is suggested by different literature that parent should buy fairly tale stories and read them for their children as much as they can, because it help them to imagine and by doing that automatically they build brain with good reasoning and analyzing capability, television watching for to long should be discouraged.

(b)Teachers/mentors role

Things that will improves child creativity;

- Teach children basing on their experiences, memories and imaginations, introduce new things/topics and let them think on their own ways their own imaginations this will discourage cramming and develop their thinking capacity. Because they will think and rethink, make those ideas and concept theirs in mind, reconstruct them, improve them and bring something new according to their level of understanding.

- Give credits to the student who do well not in terms of grades only but on the ways of his or her trying to explain on the examination; the level of detail ones gives and proofs, and examples more than taught in the class. And if student fails, teachers should asses it deeply and give a true and real feedback on time. So he or she can know the mistakes and correct them.

- As a teacher avoid limiting students to think and express their passion and creativity; limiting them using the same methods or following the same procedures when solving a certain problem. Let them use what they want to use especially in doing experiments and observation because it's from there where we can get new things, because education is not limited. The formal education are the thoughts of other people that were put together in a certain curriculum so other people can learn; good thoughts and technology and new ways of doing things in this world can come from anybody. So let the student explore their talents at times and in this way you will improve their creativity. Avoid discouraging students instead find a way to encourage them in what they are trying to do. Also as a teacher try to arrange some sort of class competition or school competitions in creativity matters from every angle; and try to award those who will come up with some amazing Idea/concepts/solutions.

- I remember our secondary school mathematics teacher Mr. Christopher Meshack used to love to say "PRACTICE PRACTICE MAKE IT PERFECT" so many times. Its true when you practice something over and over again you will be expert on that, for this reason avoids demonstrating things too much to the students and instead try to let them do. Doing something directly is best way of learning it. Because they will face some difficulties at the beginning but in the end they will manage. In doing so they will even come up with their own way of doing it due to the difficulties they face, and you will be improving their creativity.

- • As a teacher you need to be aware of what you want your student to be. Try to teach student and make them understand, don't jump to example immediately, try to define the concept and make them understand slowly, and then allow them to say something on the matter. Ask questions and let them answer according to what they have understood, this will make them use their mind and think, after that you can jump to example.

- Teachers shouldn't focus too much on the neatness of the work but rather to the creation or originality of the work. It means some people might have bad hand writings or their work is not clean but they have got very good points and very creative thoughts. Try to encourage originality, because others are able to write very well and neat but they don't have points.

- A teacher should make it clear to students on the method he is using to improve their creativity. You might give student freedom of thinking and express their talent, but in order for a teacher to get expected results the freedom has to have a certain focus and purpose to achieve in order not to go out of line. Both sides will cooperate and results will be seen in a short time.

- Try to involve your students in class in doing things, and take suggestions from them to. Don't always be the one to speak or class should not depend on everything from you, even if you know things. Also avoid giving out answers easily instead try to make student understand questions and let them get involved in the process of finding answers. Let student make mistakes because it's a part and parcel of learning. Albert Einstein Once said –Who didn't make mistake didn't learn anything in this world, so challenge your students, let them make mistakes and finally they will understand and be experts.

(c) Society role

In order for innovation to grow everybody should get involved. We may think and bring something good in our community but for it to be successful. The society must accept it and utilize. If designers create good clothes and shoes our society should buy and support. By doing this, it will encourage them because they will profit from their work. Also community should create a tradition of buying original works of the makers and not fake copy and fight against piracy. We should live in the society where we respect intellectual works of any person and intellectual work should not be free.

(d) Student role

Student has a role to play in making his innovation ability to grow and be inventive, innovative and creative, by raising critical thinking ability and training his brain to think outside the box. Student needs to puts their education into practice and also be very curious and imaginative; in doing so they will raise their problem solving ability. Albert Einstein once said I am not genius I am just curious and I stay with problems for so long to find solutions.

(e) Government role

As I explained in chapter one, Government has a huge role to play in the development of our nation,

Government should make good laws which are in favors' of us innovators and harsh to those who violate them. On top of that, all these laws should be well enforced. For them to be active government need to be extra serious.

More investment should be done by government in supporting research and development. Research is so vital in the development of science and technology. Government should put more money in this sector; we have seen the effect of investing in R&D in big and developed countries like GERMANY, USA, CHINA, and JAPAN. Also in big companies like Microsoft, Apple, Samsung and Huawei. They spend billions of dollars to prepare the future technologies.

I foresee the future"

The future is bright for those who get prepared and those who work hard and believe in it, I am among those who foresee the future of my country.

But in order to get there, we need to prepare some of basic principles.

- We should focus on invention/creation/innovation work

- We should focus on doing researches and make use of the result from it. I have seen a couple of Institutes and scientists in the country who are doing number of researches for a number of years in this country but the implementation of what they come up with are not taken seriously.

That's a major hold back factor to most African countries including Tanzania. The government should emphasize on this because a lot of money are being injected.

It is a loss and wastage of money and time to do something that you will not use. Also much emphasis for the recognition and encouragement to people who dedicate a lot of their time and knowledge to try to find solutions for various problems.

Currently I have seen that government and private sector as well as society showing much support in Music and Movie Industry which is a very good thing because our local artistes are being encouraged. Also they get publicity, shows and money. In few years to come we will witness a big change in those industries which will totally change the lives of many artists.

By saying this, I see the important development in different areas of art, including innovation, Invention and creation. To get enough exposure and support from the media, government and society in general, we will witness more wonders in the future.

Let us live our life with a dream/purpose/goal; but if you don't know how to dream make sure at least you positively affect a life of someone out there. In doing so, you will leave a mark/legacy/contribution fulfilling the purpose of living. You might not be seen by majority but by whom you change his life; you will be considered as hero and legend. Let's think as geniuses and act like innovators in order to make a difference in our communities. Also making our life more meaningful here on earth.

INDEX

INDEX

REFERENCES

Tanzania Copyright and neighboring rights, 1999
CRC-rep working paper cw001
Baum, (2004), ceo of dial,
Prather (2010)
Hansen & birkinshaw (2007
Hamel & getz (2004)
Wal-mart company-(from www.moneynewsnow.com)
Amazon.com (from marketingplan.net)
Virgin atlantic airlines-(bussinesscasestudies.co.uk)
Invention of bulb-(wikipedia)
Ayn rand, philosophy, which needs it (p. 2)
Hove, genal m
 mendelman (2007
(pescatore, 2007, p. 330).
(pescatore, 2007, pp. 336-337).
(paul & elder, 2008a, p. 88).
Hayes and devitt (2008)
(Piscatorial, 2007, p. 330).
And life success (p. 51). Ketelhut, nelson, clark, and dede (2010
Paul and elder (2009a) (p. 326). Pittman (2010)
Pescatore (2007
Literature review of idea management) anna rose vagn jensen,2012.
Davinci site-how billgates got rich
Wikipedia

ABBREVIATIONS

R&D- Research and development.
CREATIVE- Creative stands for customer-focused, risk-tolerant, entrepreneurial,
and alignment with strategy, technology and scientific excellence, innovative, virtual
organization (collaboration) and execution
HTT-Helios Telecom Tanzania
NMB-National Microfinance Bank
COSOTA-Copyright society of Tanzania
COSTECH- COMMISION OF SCIENCE AND TECHNLOGY F TANZANIA
BRELA-BUSSNESS REGISTRATION LICENCING AGENCY
LTE-LONG TERM EVOLUTION

African
Inspirations
Creative
Mind
Talents
Works Daily
Use
Exersice
Inspirations Ideas
Think
Differently
Great
Dreams
Solutions
African
Invention
Great
Inventions
Dreams
Daily
Imaginations
Make
A Difference

www.ingramcontent.com/pod-product-compliance
Lightning Source LLC
Chambersburg PA
CBHW041146210326
41519CB00046B/139